Fuel Cell Projects for the Evil Genius

Dmitry Liapitch - Instructor

Evil Genius Series

Fuel Cell Projects for the Evil Genius

Gavin D. J. Harper

New York   Chicago   San Francisco   Lisbon   London   Madrid
Mexico City   Milan   New Delhi   San Juan   Seoul
Singapore   Sydney   Toronto

Library of Congress Cataloging-in-Publication Data

Harper, Gavin D. J.

   Fuel cell projects for the evil genius / Gavin D. J. Harper.

    p. cm – (Evil genius series)

   ISBN 978-0-07-149659-9 (alk. paper)

  1. Fuel cells. I. Title.

   TK2931.H35 2006

   621. 31′2429–dc22   2008008028

McGraw-Hill books are available at special quantity discounts to use as premiums and sales promotions, or for use in corporate training programs. To contact a representative please visit the Contact Us pages at www.mhprofessional.com.

Fuel Cell Projects for the Evil Genius

This book is printed on acid-free paper.

Sponsoring Editor
Judy Bass

Acquisitions Coordinator
Rebecca Behrens

Editing Supervisor
David E. Fogarty

Project Manager
Andy Baxter

Copy Editor
Sharon Cawood

Indexer
Golden Paradox

Production Supervisor
Pamela A. Pelton

Composition
Keyword Group Ltd.

Art Director, Cover
Jeff Weeks

About the Author

Gavin D.J. Harper is the author of *50 Awesome Auto Projects for the Evil Genius, Build Your Own Car PC, 50 Model Rocket Projects for the Evil Genius,* and *Solar Energy Projects for the Evil Genius* (all from McGraw-Hill), and has had work mentioned in the journal *Science*.

Gavin holds a Diploma in Design & Innovation and a Bachelor of Science (Honours) Degree in Technology from the Open University, UK. He went on to study toward a Master of Science in Architecture with Advanced Environmental & Energy Studies with the University of East London at the Centre for Alternative Technology. He also holds the Diploma of Vilnius University, Lithuania. He has undertaken further study with the Open University and with Loughborough University's Centre for Renewable Energy Systems Technology. He is currently reading for his Ph.D. into the impacts of alternative vehicles and fuels at Cardiff University, Wales.

To John, Helen and baby Ella Maeve,
The Bullas Clan

Contents

Foreword

Global warming, depleting oil supplies, and energy independence are all topics of discussion at dinner tables around the world. Politicians argue about how quickly we need to move to renewable energy, what the interim steps might be, and how we will fund the necessary new technology. Hydrogen is the most abundant, efficient energy storage device in the universe. Converting that stored energy to electricity will be one of the ways we will overcome some parts of the looming energy crisis. However, few people have an understanding of how fuel cells *actually* work. I have anxiously awaited the printing of *Fuel Cell Projects for the Evil Genius*, so I can share it with teachers, my friends, and colleagues around the world.

Victoria Station of the London Underground might seem an unlikely beginning point for a book on hydrogen fuel cell experiments. But it was there that I met Gavin Harper with his arms full of an assortment of fuel cell equipment and a look of adventure on his face. Gavin graciously guided me through the streets of London as we searched for an elusive bottle of distilled water so that I could tutor him in the finer points of some of the small cells I had sent him from Fuel Cell Store. Gavin was determined to do something about the energy crisis in his unique and meaningful fashion. We discussed the infant fuel cell industry, the energy crisis, and, a bit off topic, Willie Nelson. I was immediately struck by Gavin's sense of wonder about the world, and encouraged by the enthusiasm and passion with which he was approaching his task of writing a book about fuel cell experiments

for students in the classroom, for hobbyists, and for the everyday person sitting at the kitchen table. It was clear that Gavin Harper was just the person to write this book. I invited him to come and work in our small laboratory in our back offices in Boulder, Colorado where he would have access to more equipment and some technical assistance.

Once there, Gavin greeted us daily, showing his cheery disposition and with frequent exclamations of joy as he found another unique way to work with hydrogen fuel cells. We were all a bit alarmed when he came out of the back one afternoon asking for a Band-Aid® because he was sure he could make a Band-Aid® methanol fuel cell. My staff and I gave the "evil genius" a Band-Aid® and sent him on his way with a wink and a nudge. A few minutes later, he came bounding from the back again—this time, to our astonishment, with a Band-Aid® methanol fuel cell running a small fan. It is with this amazing sense of wonder, and a delightful sense of humor combined with a clear understanding of the workings of fuel cells that *Fuel Cell Projects for the Evil Genius* has been written. Enjoy—play—discover the future of hydrogen fuel cell energy for yourself, guided by Gavin Harper's imagination and wisdom.

Kathleen (Kay) Quinn Larson

Founder/Director of Marketing

Fuel Cell Store

Acknowledgments

This book would not have come into being without the generous support of Kay Larson of Fuel Cell Store. I am overwhelmed by Kay's unceasing devotion to promoting fuel cell technologies to young people. You'll see her at many fuel cell conferences, and if you take part in the International Youth Fuel Cell Competition, you are likely to bump into Kay.

Also from Fuel Cell Store, I'd like to thank Quinn Larson, Matt Flood, Jason Burch, and last, but by no means least, Brett Holland for their support and encouragement. It would also be very wrong not to mention H_2 the cat—a fantastic companion to a week of fuel-cell experimentation!

I also owe a great debt of gratitude to the guys at the PURE Energy Centre in Unst, in the Shetland Islands—Dr Daniel Aklil D'Halluin, Ross Gazey, Elizabeth Johnson, and Laura Stewart. PURE are leading the way in the UK with renewable fuel cell education.

It's great to visit PURE and see the technology that we are talking about in this book being implemented in the real world "on a desktop scale"—demonstration projects like PURE provide a great opportunity for those with an interest to go and see this technology "in the metal," providing a glimpse into the future.

I'd also like to thank Allan Jones MBE for his time and theosophy on fuel cells, all of which has helped to shape my view that hydrogen and fuel cell technology will enable rapid change in our energy infrastructure.

Thanks also to Tony Marmont for doing such a great job of stimulating research and scholarly discourse regarding clean technologies in the UK, and for giving renewable hydrogen a real boost with his "Hydrogen and Renewables Integration" project. Thank you to Rupert Gammon of Bryte Energy, the system architect for the HARI project, for sharing his insight about the project.

Finally, these books would never come into the world were it not for the marvelous folks behind the scenes who make them happen. I am, as ever, indebted to Judy Bass, for giving me the chance to write, providing encouragement, and being tolerant of my transgressions. Thanks are also due to the fabulous folk at Keyword, notably Andy Baxter, for taking my text and pictures, and turning them into the book in front of you.

Chapter 1

History of Hydrogen and Fuel Cells

The way that the "hydrogen economy" and fuel cells seem to be "ahead of the curve," and "tomorrow's technology," you would be forgiven for thinking that they are a recent innovation; however, our story starts over 500 years ago.

The history of the fuel cell starts a little while before the development of the fuel cell itself. In this chapter, we are going to look at the history of hydrogen and the fuel cell, which puts the projects we are going to do in the rest of this book into context. Know that as you work through the projects in the book, you are following in the footsteps of some of the great scientists and engineers mentioned in this chapter. Become a proficient fuel cell scientist and your name could well be added to this list in years to come!

The history of hydrogen

The latin name for hydrogen is "hydrogenium," which comes from the Greek words *hydro* meaning water, and *genes* which means forming.

The first documented production of hydrogen was carried out by Theophratus Bombastus von Hohenheim (1493–1591). As this name was a bit of a mouthful, he was largely known by the name *Paracelsus*—para, meaning higher than, and celsus being the name of a very intelligent chap, Aulus Cornelius Celsus, who lived in the first century—modesty's a great character trait!

He produced a gas by reacting metals with acids, and we will see a little bit about that in Chapter 3 on hydrogen production. He did not know what it was that he had produced, although he knew the gas was flammable.

Some years later, gentleman scientist Robert Boyle (1627–1691), best known for Boyle's Law (which sounds like a dodgy a 1970s cop series, but is in fact one of the "Gas Laws" that we will explore later in Chapter 4 on hydrogen storage) looked at the reaction between iron filings and acids, rediscovering Paracelsus' earlier experiment. He published his findings in a paper: "New experiments touching the relation betwixt flame and air" in 1671. He called the gas that he produced "inflammable solution of Mars."

However, the first person to recognize hydrogen as a substance in its own right was Henry Cavendish (1731–1810), who in 1766 noted that the gas was flammable, and that it produced water

Figure 1-1 *Paracelsus—discovered hydrogen, but wasn't sure what he had discovered.*

Figure 1-2   *Robert Boyle—discovered hydrogen . . . again.*

Figure 1-3   *Henry Cavendish—decided hydrogen was a unique substance.*

when it burned. Cavendish called the gas "flammable air" and published his findings in a paper titled "On Factitious Airs." For his endeavors, he received the Royal Society's Copley Medal.

The leap that Cavendish made was to identify hydrogen as being different from any other gas. He did this by carefully measuring the density of hydrogen, and also looking at how much was produced when different amounts of metal and acid reacted. This systematic, thorough inquiry has earned Cavendish the title of the person who ultimately "discovered" hydrogen; however, this "flammable gas" was still without a name.

Jacques Charles decided to employ hydrogen and put it to good use. This is the first time that we see hydrogen in a transport application.

In 1783, ten days after the Montgolfier brothers rose into the air with their hot air balloon, Jacques Charles ascended into the air in his balloon to a height of 550 meters in his balloon "La Charlière." He also discovered Charles' Law which is essential

Figure 1-4   *Jacques Alexandre César Charles.*

Figure 1-5   *"La Charlière"*.

Figure 1-6   *Antoine-Laurent de Lavoisier.*

to know when looking at hydrogen storage. This states that under constant pressure, a gas's volume will be in proportion to its temperature.

Antoine Lavoisier has been called the father of modern chemistry. Among his great accomplishments, he disproved the "phlogiston theory," an obsolete scientific theory that held that in addition to what were then thought of as the four elements—an idea dating back to the ancient Greeks—there was an element called *phlogiston* which was held in flammable objects, and released in varying amounts as things burned.

Lavoisier also discovered that the constituents of water were hydrogen and oxygen, and went a long way towards discovering the composition of the gases that make up air. He found that when hydrogen burned in the presence of oxygen, dew was formed. This built upon earlier observations

by Joseph Priestley. Lavoisier gave the gas the name *hydrogen*, which in Greek translates to water-forming.

The science of chemistry took a dramatic leap forward with the work of John Dalton. Dalton proposed a theory of atomic structure, and devised the first table of chemical elements, where elements were assigned relative atomic weights. This was published in his work, *A New System of Chemical Philosophy*. Far from using the common letter symbols which are used to in modern-day chemistry, Dalton devised a series of graphical symbols (as seen in Figure 1-9a).

Dalton devised the symbol for hydrogen consisting of a circle enclosing a central dot (Figure 1-9b).

Now we use the symbol "H," with a small subscript "2," signifying that hydrogen naturally occurs in a diatomic state. The Russian scientist Dmitri Mendeleev (Figure 1-10), made a presentation to the Russian Chemical Society,

Figure 1-7   *Lavoisier's apparatus for his hydrogen combustion experiment.*

Figure 1-8   *John Dalton*

ELEMENTS

⊙	Hydrogen	7	⊖	Strontian	45
⊘	Azote	5	⊕	Barytes	61
●	Carbon	51	Ⓘ	Iron	50
○	Oxygen	7	Ⓩ	Zinc	56
⊘	Phosphorus	9	Ⓒ	Copper	56
⊕	Sulphur	13	Ⓛ	Lead	90
⊗	Magnesia	20	Ⓢ	Silver	190
⊗	Lime	24	⊙	Gold	190
⊖	Soda	28	Ⓟ	Platina	190
⊖	Potash	41	⊗	Mercury	161

Figure 1-9a   *John Dalton's elements in "A New System Of Chemical Philosophy" (1808).*

Figure 1-9b *Dalton's symbol for hydrogen.*

entitled *"The Dependence between the Properties of the Atomic Weights of the Elements,"* in which he arranged the elements according to their atomic mass. In Mendeleev's periodic table, hydrogen adopts the symbol H.

Now we move on from the history of hydrogen, to the history of the fuel cell.

History of the fuel cell

The recent flurry of activity by the scientific and engineering community has brought the fuel cell into the public eye. Many would believe that the fuel cell was a recent innovation, however, its roots can be traced back to as early as 1838.

Fuel cells are believed to be a very modern, current technology by many people: however, their origins can be traced back to the work of Sir William Robert Grove (Figure 1-11).

Sir William Robert Grove is widely heralded as the "Father of the Fuel Cell." He was born in 1811, in Swansea, Wales. A Welsh lawyer who later applied himself to the mastery of science, he published a diagram in 1843 and made a primitive model known as the "Grove Gas Battery."

In the journal, *Electrochemistry: History and Theory*, published in 1896, Wilhelm Ostwald described Grove's gas battery (Figure 1-12) as "of no practical importance but quite significant for its theoretical interest."

The Swiss scientist Christian Friedrich Schönbein can be credited with discovering the underpinning principles behind the fuel cell's operation. There were two schools of thought: the "contact" theory and the "chemical" theory. The contact theory stated that physical contact between materials generated the electricity, while the chemical theory stated that it was chemical reactions which generated the electricity. A furious debate ensued in the scientific

Figure 1-10 *Dmitri Ivanovich Mendeleev—*
Дмитрий Иванович Менделеев

Figure 1-11 *Sir William Robert Grove.*

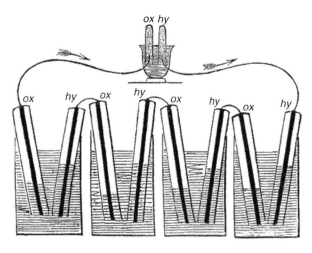

Figure 1-12 *The Grove Gas Battery.*

Figure 1-14 *Francis Thomas Bacon.*

community, and Grove's correspondent Schönbein argued for the chemical theory.

The idea of fuel cells never really developed beyond a laboratory curiosity after this discovery; it was not really until much later, in 1959, that a fuel cell with a sizable power output—5 kW—was developed by British engineer, Francis Thomas Bacon (Figure 1-14).

Bacon's fuel cell used a different technology than Grove's earlier design. He used electrodes of nickel in a solution of potassium hydroxide, initially calling his invention the "Bacon Cell."

We now recognize this family of fuel cells as "alkaline fuel cells"—there is a chapter all about them later in this book (Chapter 6)!

The first fuel cell vehicle can probably be accredited to Harry Karl Ihrig, who built a 15 kW fuel cell tractor in 1959 (shown in Figure 1-15) for Allis-Chalmers. The alkaline fuel cell used had a potassium hydroxide electrolyte and produced 20 hp power.

General Electric made a bold leap forward in the early 1960s by developing the proton exchange membrane fuel cell. This was used in the U.S. space program in the Gemini V. The PEM fuel cell was based on a Teflon™ (the material that you

Figure 1-13 *Christian Friedrich Schönbein.*

Figure 1-15 *Harry Ihrig's fuel cell tractor.*

find on your nonstick frying pans) solid electrolyte, which was impregnated with acid.

The Apollo, Apollo–Soyuz, Skylab and Shuttle reverted back to Bacon's design of an alkaline fuel cell.

Since these early days of the fuel cell the number of fuel cell *technologies* have grown, and our understanding of the material science of fuel cells has improved rapidly.

A number of technology application prototypes were developed in the 1960s by major vehicle and fuel manufacturers, to demonstrate how fuel cells could be used to provide power for transport applications. However, the 1960s was a time of cheap, plentiful oil when the world was not looking for a solution to future problems. The energy density

Figure 1-17 *The alkaline "Bacon Cells" used in the Apollo missions. Image courtesy NASA.*

of early fuel cells was not sufficient to allow practical vehicles to be manufactured beyond prototypes and demonstration vehicles.

In the beginning of the 1970s, there was a renewed interest in fuel cells, driven by the wider concern over vehicle emissions, and the damage that vehicles were doing to the environment.

The oil shocks of the early 1970s highlighted to the international community the danger of over-reliance on oil as an energy source, and people began to investigate the alternatives.

Sustained development by vehicle manufacturers over the course of the 1980s meant that the internal combustion engine's efficiency and emissions improved significantly to the point where it continued to be a viable solution. Many improvements were made as a result of the increased control over internal combustion engines afforded by new electronic engine management systems, and lean-burn engines reduced many harmful emissions. However, improvements were incremental and the fundamental technology underpinning our vehicles changed relatively little.

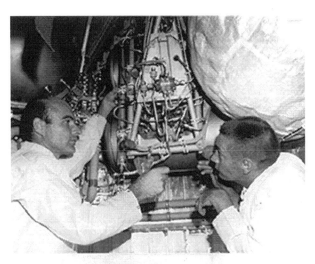

Figure 1-16 *Proton exchange membrane fuel cell in Gemini V, 1965. Image courtesy NASA.*

In 1993, Ballard Power Systems exhibited a fuel cell bus, which made important improvements on previous fuel cell vehicle applications by showing that higher power densities could be achieved with fuel cell technology. This has prompted a renaissance of major interest in fuel cell technologies. In the last decade, with an interest in cleaner, greener ways of doing things, driven by concerns over peak oil, energy security, and climate change, fuel cells have again been in the spotlight as a potential solution to some of our energy dilemmas.

In 2003, the first public hydrogen filling station was opened in Reykjavik, Iceland, which serves the three hydrogen buses in the city as part of the CUTE project.

With improvements in the energy density of fuel cells, it is not only the big vehicles that can now use fuel cell technology, but also much smaller modes of transportation. In 2005, Intelligent Energy released the first fuel cell motorcycle, capable of travelling up to 50 mph, and with a range of 100 miles in an urban driving cycle. It really doesn't make sense to transport around several tons of "dead weight" for one-passenger journey—with questions over fuel security, we may have to adapt, and realize that some of the most sensible solutions to problems aren't technical ones, but societal ones—driving small vehicles more fitting for our needs than our wants.

So what does the near future hold for fuel cells?

There are a number of fuel cell projects that are close to completion, and appear to be technologically feasible in the near term. We can already see a wide range of fuel cell applications that have emerged in the past couple of years; and with the experience and information learned from demonstration projects, the next generation of demonstrations will have improved upon the current state of the art by employing the knowledge gained. While there is further work to be done in the labs by the "materials scientists" working on improving the chemical process to convert fuel to power and to make fuel cells longer lasting, cheaper, and more

reliable, it is important that we evil geniuses do not lose sight of the fact that by creating new, novel applications for fuel cells, we encourage investment in the fuel cell marketplace, which brings the technology closer, quicker.

Fuel cells are such a cutting edge technology that applications are being developed for them every week—not just by big, well-funded industry, but equally by teams of college or university students. A team called Energy Quest plans to race their hydrogen fuel cell powered boat, Triton, in the near future.

Howaldtswerke-Deutsche Werft AG (HDW), a German company, has developed the Class 212 submarine, a nonnuclear submarine that can remain submerged for weeks, without needing to come to the surface because of its PEM fuel cell technology.

WWW.

If you are interested in the history of fuel cells, or maybe even need more information for a school project or report, there is a rich archive of information on the Internet containing further reading, on not just the *low-temperature* fuel cells that we will be experimenting with in this book, but also the *high-temperature* fuel cells, which because of their high operating temperatures aren't suitable for home experimentation.

The Smithsonians' history of fuel cells, comprehensively covers the higher temperature fuel cells which are not included in this book:

americanhistory.si.edu/fuelcells/

Fuel Cell Store's history of fuel cells:

www.fuelcellstore.com/information/fuel_cell_history.html

A history of fuel cell development at Los Alamos National Laboratory:

www.lanl.gov/orgs/mpa/mpa11/history.htm

Chapter 2

The Hydrogen Economy

You've heard a little bit so far about the history of hydrogen, and how fuel cells came into being, but why are we so interested in this gas? What is so special about it? Before we launch into the experiments, you need to be aware of why we are doing this, and why fuel cells are the next big thing for smart evil geniuses to be interested in. We are going to explore the physical properties of hydrogen gas, and look at how—with fuel cell technology—hydrogen could become a player in meeting our future energy needs, and delivering the sustainable energy revolution.

Hydrogen

Take a look at any periodic table, and you will find hydrogen right at the top. It is element "numero uno," the first element in the table and the lightest of all elements. The gas is colorless, tasteless (although we don't recommend tasting chemicals!), and nonmetallic. If you take a peek at the periodic table, you will see hydrogen in the first box to the top left (just above lithium). The box that you will see will look something like Figure 2-1.

You will see at the top of the box the number "1"—this is hydrogen's place in the periodic table. Beneath the "1"—you'll see "H," the chemical symbol for hydrogen, and beneath this, you'll see 1.00794 this is how many grams a "mole" of hydrogen weighs, which makes it the lightest element.

Hydrogen comprises of one "proton," and one "electron" in its most common form "protium"—hydrogen is a "diatomic" molecule. This means that naturally occurring hydrogen is

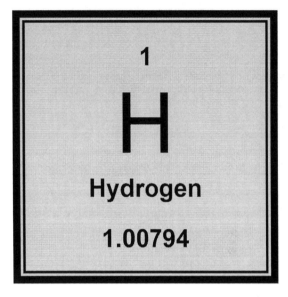

Figure 2-1 *Hydrogen as it appears in the periodic table.*

comprized of two atoms of hydrogen (for those of you who don't know Greek, *di*-is a prefix meaning "two").

Looking at Figure 2-2, we can see how there are two atoms that make up diatomic hydrogen. The protons in diatomic hydrogen occupy the two black dots that represent the central nuclei, while the electrons can be in any position in a "cloud" surrounding them. This is because we cannot say with certainty *where* exactly the electron will be,

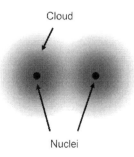

Figure 2-2 *Representation of diatomic hydrogen.*

but we know the probability of it being a certain distance from the nucleus. The gradient "probability cloud" illustrates this.

At room temperature, hydrogen is a gas; in fact, it boils at 20.28 Kelvin which is the same as −252.87° Celsius, or −423.17° Fahrenheit.

One of the problems is in trying to store this hydrogen in a form that is compact—because even though hydrogen doesn't weigh much, in its gaseous form, it takes up a very large volume! Chapter 4 will be looking at different ways of storing hydrogen.

To give you an idea, of the difference in volume between liquid hydrogen and gaseous hydrogen, take a look at Figure 2-3. Hydrogen has an expansion ratio of 1:848. This means that when it is stored as a liquid, a volume of "1 unit" of gas will increase 848 times when allowed to expand into a gaseous state.

Hydrogen is the most abundant element in the universe, however, on Earth, there are only 0.55 parts per million by volume of hydrogen in our atmosphere. Most of the hydrogen on Earth is locked up in the water that makes up our seas, lakes, and rivers. The chemical formula for water is H_2O, which means that water contains two hydrogen atoms and one oxygen atom.

Our fuel cells take hydrogen fuel, and oxygen from the air, and produce water, releasing energy

in the process. In the process of doing so, they release energy—"chemical energy." Where does this energy come from?

Hydrogen and oxygen, as elements, can be considered to be at a *high* energy state. When combined together as water, they can be considered to have a *low* energy state. In the process of combining them—whether by combustion or using a fuel cell—the chemicals change from a *high* to a *low* energy state, but this energy must go somewhere! (As energy is never created or destroyed—only changed from one form to another). This energy is the energy that manifests itself as heat if we burn the hydrogen, or as electrical energy if we use the hydrogen in a fuel cell. This change in energy states is illustrated in Figure 2-4.

Now, we need to get hydrogen from somewhere. In the same way we can make water with hydrogen and oxygen, we can also make hydrogen and oxygen from water; however, we also need to "input" energy to disassociate hydrogen from oxygen in the process of electrolysis.

Electrolysis is one of the ways that we can produce hydrogen, and we will carry out some electrolysis experiments later in this book.

Hydrogen has the potential to be a very clean, green, fuel. In fact, it's very easy to get into the habit of calling it a "fuel"—after all, we use it in

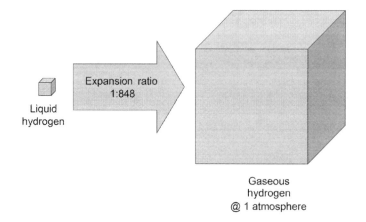

Figure 2-3 *The expansion ratio of hydrogen.*

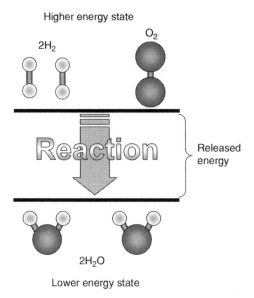

Figure 2-4 *Hydrogen and oxygen combine to make water, releasing energy in the process!*

"fuel cells." However, since we can't easily get hydrogen in its elemental form—we can't dig it up from the ground—it is better thought of as an "energy carrier," a bit like electricity is a "carrier" for energy, but not really a fuel.

So, hydrogen can be a "carrier" for "clean energy" or it can be a carrier for "dirty energy."

We need to make an *input* of energy somewhere into the process to produce hydrogen. The only emissions where the hydrogen is consumed is pure, clean water—no problems so far, but problems can arise if the energy we use to produce the hydrogen comes from "dirty" sources.

"Green" hydrogen

So, there are a number of ways in which we can produce "green" hydrogen. We can use renewable energy technologies to produce electricity, and this electricity can then in turn be used to produce hydrogen gas by a process called electrolysis. We can see this process in Figure 2-5.

Let's explore some of these different energy technologies in a little more detail.

Solar power

We can harness the power of the sun in photovoltaic cells. In fact, we can produce electricity using photovoltaic cells and feed this to a separate

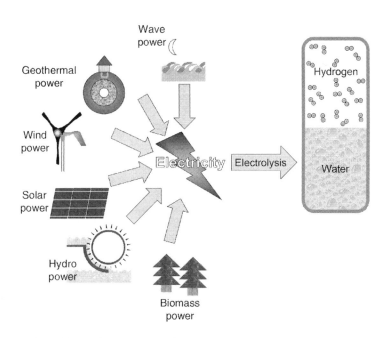

Figure 2-5 *Hydrogen produced from clean, renewable energy.*

Figure 2-6 *A solar array showing solar power—energy from the sun.*

electrolyzer, or, we can use the photovoltaic cells effectively "immersed in water," to electrolyze the water directly. Some scientists are working on types of algae that can harness the sun's energy directly to produce hydrogen. All of these processes use the sun, which is clean, free and cheap—our safest nuclear reactor—9.295×10^7 miles away from the Earth, or if you are travelling in a hydrogen-fueled spaceship going at lightspeed (not yet invented, so watch this space), you could be there in 8.31 minutes.

Note: If you are interested in learning more about solar energy, you might like to check out my other publication, *Solar Energy Projects for the Evil Genius* (McGraw-Hill).

Wind power

We can also harness the power of the wind, using wind turbines, to produce electricity, which can then be used to produce hydrogen cleanly. Turbines take the movement of the air (which is caused by air rushing from an area of high pressure to an area of low pressure) and use this force to turn the blades of a turbine. As the turbine blades turn, they

Figure 2-7 *A wind turbine illustrating the power we can harness from the wind.*

drive a generator; sometimes directly, sometimes through a gearbox. The power from this generator is then rectified. Once we have this electricity, we can feed it into the electricity grid that supplies our homes, and then use the power to electrolyze water into hydrogen at times when the wind is blowing, but people aren't using much electricity (as in the still of night when everyone is in bed). This makes use of cheap energy to produce our hydrogen. Additionally, in remote islands or isolated locations, hydrogen can be used to store electricity and even out fluctuations in load, even when there is no connection to the electricity grid.

Hydro power

The hydrological cycle is where water evaporates from the seas and the land, driven by the sun, collects in the clouds, and rains back to the ground (a process called precipitation). When the rain lands on the ground, it can land on mountains or hills, (high areas) where it must then, under the influence of gravity, make its way back to the seas. This forms rivers and streams all over the land. We can *capture* water when it is at a high point, channel it through a turbine or waterwheel, and capture energy from the turbine to produce electricity. In the process of falling from a *high* location to a *low* location, the water releases *gravitational potential energy*, and it is this that we capture in a hydrostation. This hydrostation can be used to electrolyze water into hydrogen and oxygen.

Wave/tidal power

Wave and tidal power are distinctly different; however, we often class them together as they are "power from the sea." Tidal power is a process that is driven by the moon as it orbits the Earth. The influence of the moon's gravity on our seas causes the water to "bulge"—this forms the tides which are regular and predictable. Mariners and seafarers will know to a great degree of accuracy when the tide is *low* or *high*, and we can use devices submerged in the sea to capture the mass of energy

Figure 2-8 *Power can be captured from falling water to generate energy.*

Figure 2-9 *Model of a wave energy capture device.*

as water rushes from place to place. Wave power, by contrast, is the power in the surface of the waves—think of surfers riding along waves, and the energy and power that waves can exert. Devices that float on the sea or near to the surface can capture these waves to turn the power into electrical energy and transmit it back to the shore. In both these instances, when the power reaches the shore, it can be used to electrolyze water into hydrogen and oxygen.

Geothermal power

Geothermal energy, energy from the heat in the core of the Earth, can be used to raise steam to drive conventional steam turbines to produce electricity, which can be used to provide hydrogen.

Biomass

All the living things that we are surrounded by—the trees and plants and crops that grow in farmers' fields—are a mass of biological matter. The sun drives a process called photosynthesis in plants,

Figure 2-10 *A plume of hot water, driven out of the ground by geothermal energy.*

which enables them to take carbon dioxide out of the air, combine it with hydrogen and oxygen from water to make food to nourish the plant, and allow it to grow. We can take biomass, and turn it into fuels—biodiesel, bioethanol, but we can also extract the hydrogen directly from biomass, such as the woodchip waste shown in Figure 2-11, using

Figure 2-11 *Biomass—power from plants.*

gasification and reformation processes, similar to those used to extract hydrogen from fossil fuels. This does produce carbon dioxide; however, the only carbon dioxide produced is that which has been extracted from the atmosphere by the plant as it grew!

"Dirty" hydrogen

At the moment, most of our hydrogen is produced using "dirty" methods. Fossil fuels, coal, oil, and natural gas, are taken and "reformed" into hydrogen. However, fossil fuels also contain a high proportion of carbon. This carbon has to go somewhere—as part of the reformation process, the byproduct, "carbon dioxide," is produced. Carbon dioxide is a greenhouse gas which adds to the "greenhouse effect," which leads to the problems that we are beginning to witness with global warming and climate change.

Coal, oil, and gas can all be taken and turned into hydrogen using a process called steam reformation, which takes the fossil fuels, combines them with steam at pressure and very high temperature, stripping the carbon, producing carbon dioxide, and capturing the hydrogen for use in fuel cell vehicles. Natural gas contains the highest ratio of hydrogen to carbon. If you think about methane

and its chemical structure (CH_4), there are four hydrogen atoms for every carbon atom. Look at oil, and the amount of carbon relative to hydrogen increases—take petrol, which contains many different lengths of hydrocarbon chain—but we can approximate to octane, and you see the ratio of hydrogen to carbon is nearer: 1:2 (C_8H_{18}) while methane is 1:4. Start to look at coal, and the story begins to get even more grim—as coal is largely comprized of carbon, there is *much* more carbon dioxide produced than hydrogen.

Alternatively, we can burn fossil fuels in a conventional fossil fuel power station, like the one in Figure 2-13, and use that electricity for electrolysis.

However, although this is very undesirable, it is not all totally bad news. Compare producing hydrogen in this way to burning the fuels directly in your vehicles and over a range of scattered locations. By producing hydrogen, you are making all of the carbon dioxide in a central location, where it can be captured. Once we have separated and captured the carbon dioxide, we don't just have to release it into the atmosphere; we can do clever things with it—like trap it, and contain it. A process called carbon sequestration can be used to take carbon dioxide, and inject it deep underground, into geological features, where the CO_2 is trapped and (touch wood) cannot escape.

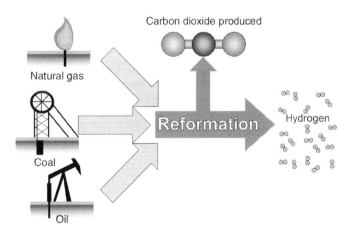

Figure 2-12   *Hydrogen produced from dirty fossil fuels.*

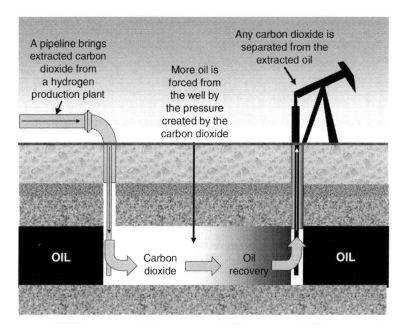

Figure 2-13   *Fossil fuel power stations can generate hydrogen to produce electricity.*

The large oil and energy interests find this technology *very* attractive. The reason for this has less to do with altruism and the preservation of the human race, and more to do with oil and profits.

Figure 2-14   *The products of combustion from fossil fuel plants like this could be captured and sequestered.*

By forcing carbon dioxide into oil wells which are nearing the end of their economic life, the carbon dioxide can be used to force additional oil to the surface, recovering deposits of oil that would otherwise go to waste.

Nuclear hydrogen

Some people advocate a hydrogen economy based upon hydrogen produced en masse by nuclear power. We can see by looking at Figure 2-15 that nuclear power is used to produce electricity, which in turn is used to convert water to hydrogen by electrolysis.

Unfortunately, the byproduct of this process is toxic, radioactive nuclear waste, which takes thousands of years to decay to safe levels—and we still haven't worked out a permanent solution for what to do with it!

The reason that some people advocate nuclear power is that no carbon emissions are produced by the actual process of producing electricity from nuclear fuel. However, take into consideration the massive amounts of concrete and building

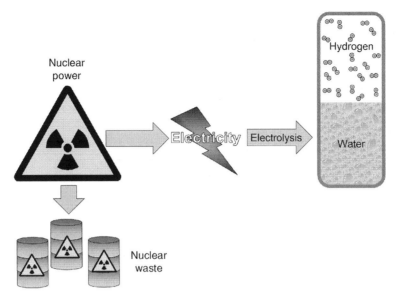

Figure 2-15   *Hydrogen produced from nuclear power.*

materials needed to build a nuclear plant, consider the carbon dioxide produced in blasting, mining, extracting, refining, transporting, storing, and processing nuclear fuel, and suddenly you realize that there is an awful lot of *hidden* carbon dioxide in this process.

Furthermore, once their useful life has finished, nuclear power stations, like the one in Figure 2-16, need to be decommissioned—a process which requires a great amount of money, and a great investment of energy—with its associated carbon emissions.

Figure 2-16   *Nuclear power stations leave an ugly blot on the landscape.*

The need for cleaner technologies

Since the Industrial Revolution, through the oil boom of the 1960s, the human race's burgeoning need for energy has resulted in a rise of carbon dioxide emissions globally.

A monitoring station on the Island of Mauna Loa, in remote Hawaii, has been monitoring the levels of carbon dioxide since the mid-1950s. The results are displayed as a graph in Figure 2-17.

It is clear to see that since the 1950s, the level of carbon dioxide in our atmosphere has continually and steadily increased.

"But how do we know that there wasn't this much carbon dioxide in the atmosphere before 1955?" some people counter. Well, in addition to taking measurements of carbon dioxide from the atmosphere, it is also possible to drill great cores of ice from the Antarctic. This ice has been formed over many, many years, as snow deposited itself on top of the existing shelf, trapping little pockets of gas. We can analyze these pockets of gas to find out an abundance of data about what our climate was like at different times.

If you want to read more about ice cores, there is a comprehensive and informative article on Wikipedia: en.wikipedia.org/wiki/Ice_cores

So, why have carbon dioxide emissions recently shot up?

The growth of our economies has been founded on cheap oil, and the perception that "the world's resources are endless." As we begin to find resource shortages, we will quickly begin to realize that the Earth's resources are indeed finite.

Ever since the Industrial Revolution, our demand for fossil fuels, and other resources has grown. Looking at Figure 2-18, we can see how suddenly man-made (anthropogenic is a word you will often hear used) emissions have soared rapidly. In fact, most of these emissions were created in the past century.

Think of all the things that we can do in this century, which wouldn't have been possible last century, or even decades ago. Think about the massive amount that we use our cars. Mass production has enabled everyone in the developed world to be able to own a car should they choose. We have seen the growth in numbers of families with not just one but many cars. All these vehicles require energy to be made, and in use. Look at aviation as another growth sector. Fifty years ago, aviation was expensive, for the elite, but now it is possible to fly so incredibly cheaply that the number of flights taken has soared (literally) into the sky!

Figure 2-17 *CO$_2$ measurements taken at Mauna Loa.*

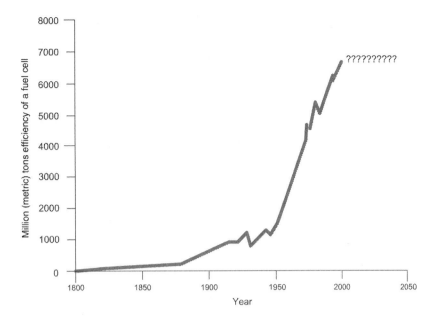

Figure 2-18 *Global fossil fuel carbon emissions since the Industrial Revolution.*

So what's the problem?

Well, the first one is climate change. If you haven't already seen Al Gore's "An Inconvenient Truth," it is a great primer on climate change for evil geniuses of all ages. The carbon dioxide that we produce when we burn fossil fuels is one of the gases that contributes to a phenomenon called "the greenhouse effect" which produces global warming.

Looking back to earlier in this chapter, you will remember that there are "clean ways" in which we can produce hydrogen, without producing any carbon dioxide emissions in the process. Of course, at the moment, our technologies and infrastructure are geared towards a *carbon-based* economy. At the moment, most of our hydrogen available for use in fuel cells comes from carbon-based fuels, however, with future development of renewable energy resources, the hydrogen economy offers some attractive prospects for a world moving toward a cleaner future.

The thing is, it isn't just the countries that are already developed that are going to require energy. While the energy demands of America and Europe continue to grow, we must also think of developing countries such as China and India, whose economies are growing *rapidly*. With this rapid growth comes increased demand on resources and energy.

And there's another reason for looking at hydrogen: you might have noticed that the price of gas at the pumps has been increasing significantly recently. Chances are that from here onward prices aren't going to come down significantly for any long period of time.

A scientist from the American Petroleum Institute, Marian King Hubbert, made a prediction in 1956 that oil discoveries would follow a bell-shaped curve. Take a look at Figure 2-19, as we are going to look through the features of the graph.

Peak oil is an important subject to read up on, for any budding evil genius interested in fuel cell technology. Wikipedia carry some good articles on this, which I highly recommend:

en.wikipedia.org/wiki/Peak_oil

en.wikipedia.org/wiki/Hubbert_peak_theory

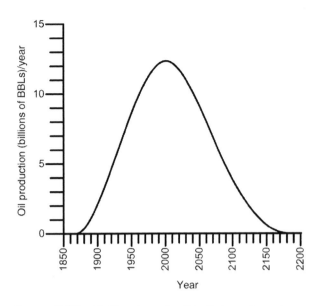

Figure 2-19 *Hubbert's peak oil bell-shaped graph.*

production—peak oil—which means from here forward, oil starts getting a lot harder to find. As a result of the increased effort that needs to be expended to find new oil, the price will begin to soar. Rather than being cheap and plentiful, oil will start to become more expensive, as it becomes harder and harder to find and produce. The price begins to rise, and rise, and rise, while the actual amount of oil entering the system begins to decline. Now, remember, we don't just need oil for transportation, we also use oil in the production of pharmaceuticals, plastics, paints, and various finishes, so we don't want to *spend* all of our budget of oil on transportation and producing energy—this could be where hydrogen and fuel cell technology enter the arena.

Hubbert said that for any geographical area—whether a single oil field, the oil reserves of a state, nation or continent—or ultimately the entire world, the rate at which oil was discovered would follow what is called a bell-shaped curve. If you look at Figure 2-19, you can see that the area enclosed beneath the line sort of resembles a crude drawing of a church bell.

Looking at the *beginning* of this curve, we see that when oil was initially discovered, it wasn't produced in great profusion, as people were still finding applications for it, and investing in infrastructure to produce and refine it. As we slowly discovered how darn useful oil actually was, people began to consume more of it, which in turn drove the production of oil. With the invention of the internal combustion engine, and the spread of the motor car, our consumption of oil has soared (with a massive increase in carbon emissions as a result). However, this growth is not without limits. There is a finite amount of oil in the Earth's crust. We have found all the oil that is "easy to discover," the oil that is in the accessible places that can be easily and cheaply reached. We are now approaching the "peak" of oil

Why hydrogen?

The ways that we currently "extract" energy from fossil fuels are pretty wasteful. If we look at Figure 2-20, we can see a range of different devices that we use for producing *useful energy* from fuels. We can see that fuel cells, over a whole range of powers, eclipse the efficiency of even the best hybrid solutions and turbine applications.

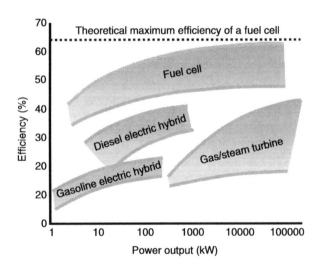

Figure 2-20 *Technologies for converting energy compared.*

The barrier at the moment is twofold:

- Technology
- Price

Fuel cells work well in the lab, but as soon as we start to take them and put them into applications, we have to overcome practical problems concerned with the technology. Our hydrogen is very light, but it takes up an enormous *volume*—as a result, we have to look at ways to practically store the hydrogen for use in our vehicles and other applications. Another is the lifespan of fuel cells. Everything in this world wears out sooner or later—the first combustion engines would have been unreliable affairs—with the technology being improved over time. With fuel cells, the technology is gradually improving. We are coming to the end of the initial "research and development" phase, and we are starting to see fuel cell technology with sufficient durability to be able to enter commercial applications.

Remember, there aren't any moving parts with hydrogen and fuel cells, so there isn't anything mechanical to go wrong. Further improvements need to be made, but step by step, the technology is getting there. For vehicles in cold climates, the cold-start performance of fuel cells has had to improve, but again this shows signs of improvement. Furthermore, manufacturers are learning how to get more power from smaller and smaller fuel cells, and we will look at some of the technology developments in this book. As fuel cells shrink, there is more room for other vehicle components, and hydrogen storage, so we see more applications for the technology. Aside from the technology, we need to look at the price. For fuel cells that employ PEM technology, platinum, the catalyst, as we will see, is a very important part of the chain of events, making the reactions in the fuel cell happen. I don't know if you've browsed any jewellers recently, but platinum is pretty expensive stuff. It really isn't cheap! This is one of the barriers to the development of fuel cells commercially, but gradually the requirement for

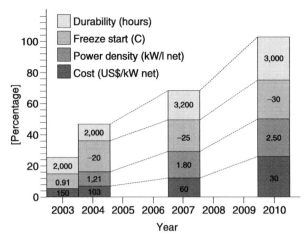

Figure 2-21 *Fuel cell technology trends in vehicles.*

platinum in fuel cells has being steadily reducing. This and other improvements to the technology will mean that fuel cells will become cheaper—and as they do so, they will become more affordable to you and I, finding more applications in the real world.

Look at Figure 2-21, and you will see some of the major technology trends for fuel cell technology in vehicles. You can see where the technology has come from, and where it is expected to go by 2010.

We looked earlier at all the different possible methods of producing hydrogen—as the technology becomes cheaper and more attractive, we can envisage an energy economy based on renewable energy and clean energy sources *generating* electricity to produce hydrogen, with this hydrogen being used as a *carrier* for the clean energy.

If you look at the Figure 2-22, you can see how a prospective hydrogen economy *could* work.

Of course, the advantage, as you can see, is that with clean energy and hydrogen, carbon dioxide and toxic pollutants are eliminated from the process, with the only things entering and exiting to and from the natural environment being water and oxygen. We can see this in Figure 2-23, hydrogen and oxygen being the constituents of water.

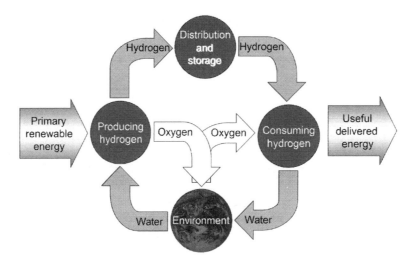

Figure 2-22 *Prospective hydrogen economy.*

Work through the projects in this book and you will begin to understand the technology, but all the time throughout your experimentation, relate the practical things that you are doing to what is happening in the *real world* outside the lab. Think about and visualize how the technology that you are working with has the potential to change the world for the better. When thinking about the technology, reading articles in newspapers and magazines, and talking to your friends and family, always keep a sound grasp of the issues—think about how clean hydrogen can be a force for good, and when you hear of hydrogen being generated from dirty sources, read between the lines and think about whether the strategies proposed really are environmentally sound.

$$2H_{2(g)} + O_{2(g)} = 2H_2O_{(aq)}$$

Figure 2-23 *Hydrogen, oxygen and water.*

Chapter 3

Producing Hydrogen

So where do we get hydrogen from?

There are a number of ways that we can get our hydrogen. It is a bit of a myth that hydrogen is a "fuel"—it isn't really as there is no such thing as a hydrogen mine or a hydrogen well—we can't just dig it out of the ground or drill for it like oil or gas. We need to produce it from another primary energy source. In this chapter we are going to evaluate some of the different options for producing hydrogen.

Electrolysis

At school, you might have used a Hoffman apparatus in science class. A Hoffman apparatus has a reservoir of water through which is passed an electric current. The electric current disassociates the hydrogen from the oxygen in the water. The gas bubbles off from the electrodes and is collected in separate storage containers. It is observed that twice as much hydrogen is produced as oxygen. Taking a little bit of time to think about this, we see that the chemical formula for water is H_2O—this makes sense as we can see that there is twice as much hydrogen in water than oxygen.

The hydrogen produced by the electrolysis process is *very pure*—some fuel cells require a very pure form of hydrogen so this is ideal.

The one disadvantage of electrolysis is that significant amounts of electrical energy are needed for the process. While this electricity can be generated using clean, green renewable energy, there are also many champions of a nuclear-hydrogen economy using supposedly *cheap* nuclear energy to produce hydrogen—this would leave us with a toxic legacy of waste, and would negate many of the benefits of a clean hydrogen economy.

Biomass gasification and reformation

Biomass has proven itself as a relatively clean, near carbon-neutral source of energy. Agricultural waste, organic matter, wood, and other sources of biomass can be heated in a controlled atmosphere without the presence of oxygen. This yields a gas—synthesis gas—which is hydrogen-rich as well as containing carbon monoxide and dioxide.

The carbon emissions from this source of energy are effectively neutral as the carbon dioxide was taken out of the atmosphere in the first place by the growing plants, however, carbon emissions in the production and distribution of biomass cannot be ignored. There is also the possibility of sequestering the carbon produced in the gasification process. This could effectively make biomass with hydrogen extraction a "carbon-negative" fuel.

Steam reformation

By combining high-temperature steam, and methane, it is possible to extract hydrogen from this fossil fuel. Although carbon dioxide is produced, the location is centralized and it opens up the possibilities for carbon sequestration (although bear in mind that this is an unproven technology with no demonstrated long-term safety). The process is fairly cheap and inexpensive, and the heat produced can also be harnessed (known as cogeneration). Cogeneration provides us with lots of low-level heat which could prove useful in local combined heat and power schemes. This method does show a lot of promise, as it is currently an efficient cheap technology that will work with the existing gas-distribution infrastructure; however, the carbon emissions are impossible to ignore.

Photoelectrolysis

Photoelectrolysis is a relatively new unproven technology. It involves using solar energy to stimulate a silicon junction similar to a photovoltaic cell, with the distinction that instead of the energy being converted to electricity, the silicon junction acts *directly* on the water where electrolysis occurs. This technology shows promise, although much development must be done.

Clean coal?

There are vast tracts of coal throughout the world; however, coal is carbon-rich—burning it doesn't help global warming, and mining leaves scars on the landscape, which can last for generations. There are, however, schemes afoot to look at gasifying coal, extracting the carbon, and sequestering it.

Biologically produced hydrogen

There are a number of types of algae that use photosynthesis to convert solar energy into hydrogen. At the moment, these processes have only been demonstrated on a small scale, but research in this area is intense. It is expected that great strides forward could be made in this area.

Project 1: Testing for Hydrogen

In this chapter, we are going to look at how to produce hydrogen. But once you've produced a gas, how do you know what it is? Here we are going to show a simple test for hydrogen.

You Will Need

- Test tube of gas under test
- Wooden splint
- Source of flame (e.g., candle)

Method

Hydrogen is a colorless, odorless gas, which is incredibly flammable, and combines with oxygen

to form water. The mixture is so explosive that just a single spark will make it explode! We can use this property to help us test for its presence!

There is a really simple, if a little loud, experiment.

Take a test tube of the gas which you believe to be hydrogen. Take a lighted splint of wood, and introduce it into the container. The hydrogen should produce a loud squeaky pop as shown in Figure 3-1.

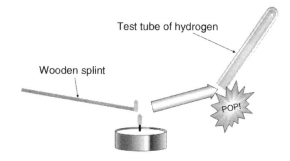
Test tube of hydrogen

Wooden splint

POP!

Figure 3-1 *The test for hydrogen.*

Check out some videos of hydrogen explosions!!!

Look at the *.mpeg "hydrogen gas"

smartweed.olemiss.edu/nmgk12/curriculum/
curr_eighth.htm

Here, a balloon full of hydrogen gas is lit!

pk014.k12.sd.us/chemistry/chapter%20two/
hydrogen_balloon.mpg

One of the things that you can do is perform this experiment near a piece of dry glass that has been chilled. You should see a frosting on the glass—this frosting is water.

Conclusion

Hydrogen is an explosive gas which will easily ignite. Its flammability is one of the reasons that we tend to use helium rather than hydrogen in novelty balloons. When hydrogen burns in air, it combines with oxygen and the product is water.

We can write this as a word equation, as in Equation 3-1

Word equation for hydrogen burning (Equation 3.1):

$$\text{hydrogen} + \text{oxygen} \rightarrow \text{water}$$

We can see that hydrogen (in our test tube), and oxygen (in the air) combine to make water (the frosting we saw on the glass).

If we want to be more sophisticated scientists, we can take this word equation, and turn it into a "chemical" equation, where we substitute the words for chemical symbols. You can see in Equation 3-2, how "H" represents hydrogen, and "O" oxygen, with the chemical formula for water H_2O. The symbols in brackets after each chemical denote the *state* of the chemical—solid, liquid, gas or dissolved in water.

Chemical equation for hydrogen burning (Equation 3-2):

$$H_2(g) + O_2(g) \rightarrow H_2O(l)$$

State symbols

Here's a handy reference of state symbols for you. Remember, we use "l" for water, as it is a liquid, but when something is *dissolved in water*, we use the symbol "(aq)" to represent "aqueous solution."

(s)	=	solid
(l)	=	liquid
(g)	=	gas
(aq)	=	aqueous solution (dissolved in water)

However, the astute mathematicians among you will quickly notice that things don't quite add up. Looking at the *amounts* of hydrogen and oxygen, and the amounts of water produced, we appear to have *lost* an oxygen atom somewhere, as there are two atoms in one molecule of hydrogen. Therefore, we need to *balance* the equation. Two lots of hydrogen will combine with one lot of oxygen to make two lots of water. We can show this by prefixing the "Hs" with a "2" to denote how many lots. This is shown in Equation 3-3.

Balanced chemical equation for hydrogen burning (Equation 3-3):

$$2H_2(g) + O_2(g) \rightarrow 2H_2O(l)$$

You are going to come across balance equations a lot in this book, so get to grips with them now. Just remember that the big number in front shows how many lots of something there are, while the little subscript numbers show how many atoms there are of an element in each lot.

Project 2: Testing for Oxygen

You Will Need

- Test tube of gas under test
- Wooden splint
- Source of flame (e.g., candle)

Method

Oxygen is the only gas that will support burning, so we can use this property to test for its presence. To test for oxygen, take a test tube of the gas that you want to test. The procedure for this test is illustrated in Figure 3-2

Ensure that the end of the tube is either sealed or immersed under water to prevent the gas from escaping. Now, take a wooden splint and light it (Step 1), blow it out (Step 2)—you should make sure that the end of the splint glows—there might be a little wisp of smoke where it is smouldering, but it should not be lit. Now dip it into the test tube

that you suspect is oxygen right away (Step 3). If the gas is oxygen, the splint will quickly relight (Step 4).

Oxygen factoid

Oxygen is pH neutral—it will not have any effect on litmus or indicator paper.

Conclusion

We have seen how oxygen is essential for burning to take place, and how burning is much easier in an environment of *pure* oxygen, than in air, which contains only 21% oxygen. This is evidenced by the fact that the splint was only *glowing* in the air; however, once inserted into the test tube of oxygen, it began to burn brightly.

We will see later on, with our fuel cells, how they operate much better with *pure* oxygen than with ordinary air.

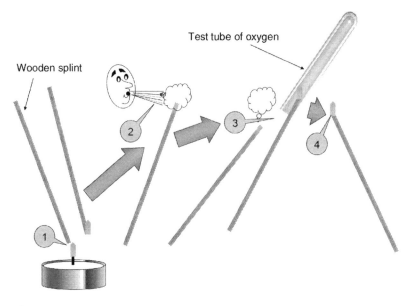

Figure 3-2 *The test for oxygen.*

Project 3: Investigating Acid–Metal Reactions

There are a lot of different ways in which we can produce hydrogen—these are going to be investigated in this chapter. One way of producing hydrogen, that is convenient using readily accessible materials in a laboratory, is by reacting an acid with metal. We are going to carry out this experiment to produce hydrogen, and then look at the chemistry that is taking place.

You Will Need

- Thistle tube
- Swan neck tube
- Gas jar
- Bowl
- Timer

Method

You are going to need to set up the apparatus as shown in Figure 3-3. Be careful with the thistle tube, swan neck tube, and flask as they are made from delicate glass.

As your reaction proceeds, use a timer and take measurements of the volume of gas produced from the marks on the gas jar. Table 3-1 has been provided for your convenience to note your results down.

Conclusion

In a reaction involving a metal, the metal reacts by losing one or more electrons. This loss of electrons is called "oxidization" by chemists; we can *rank* metals according to how easily they *oxidize*; the more easily they oxidize, the more *active* we say that they are. The activity series for metals is shown in Figure 3-6. Hydrogen is shown on this list as a reference point (although remember it is a nonmetal).

At the same time as oxidization is taking place, it is always accompanied by a *reduction* of materials. Reduction is where a chemical *gains* electrons.

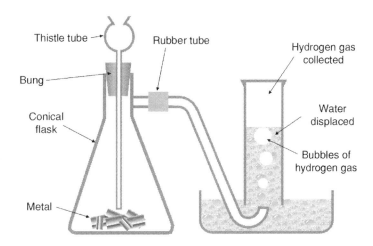

Figure 3-3 *Producing hydrogen from acid and metal.*

Table 3-1
Results from acid–metal reaction.

Time (seconds)	Hydrogen produced (cm³)
s	cm³
s	cm³
s	cm³
s	cm³
s	cm³
s	cm³
s	cm³
s	cm³
s	cm³
s	cm³

Remember that oxidation is always accompanied by reduction by using this simple little diagram.

A nice way to remember this is illustrated in Figure 3-5—this is a diagram you'll come into contact with later in this book! Remember the words LEO and GER using the simple mnemonic—LEO the Lion goes GER—Loss Equals Oxidation, Gain Equals Reduction.

Figure 3-4 *Reduction, oxidization—like Starsky and Hutch, you can't have one without the other*

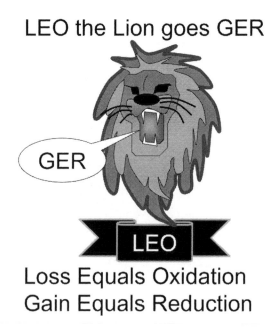

LEO the Lion goes GER

GER

LEO

Loss Equals Oxidation
Gain Equals Reduction

Figure 3-5 *LEO the Lion will help you to remember redox reactions by going GER.*

So what's happening here?

The hydrogen ions in the acid migrate to the metal surface. Hydrogen ions in the acid accept electrons from the metal surface (and the metal surface gives up electrons). The hydrogen ions combine with the electrons to make hydrogen atoms. Hydrogen atoms combine to give hydrogen molecules. Remember from Chapter 2 that with hydrogen it takes "two atoms to tango," as it is *diatomic*. Metal ions hydrate and move away from the metal surface.

You will notice that the production of hydrogen gas eventually slows down and halts. As the reactants are consumed, the rate of hydrogen production declines until all the reactants are used up—and then the reaction stops.

Taking it further

We are going to be hearing a lot in this book about catalysts, so you might as well get used to them sooner rather than later! The production of hydrogen using zinc and sulfuric acid was really quite slow. If we wanted to increase the rate of reaction, a good way to do this would be to use a *catalyst*, a chemical which *facilitates* the reaction and enables it to proceed at a faster rate.

You can add a little copper (II) sulfate to your zinc, and that will speed up the rate of hydrogen production. You should already have some data about the reaction rate of zinc. Try adding a little catalyst, and see how it affects the rate of reaction. Note your results in Table 3-2.

Compare the gas produced over time in Table 3-1 with Table 3-2 when the catalyst is added . . . Interesting?

Table 3-2

Comparing the reactions with and without catalysts.

Time (seconds)	With $CuSO_4$ catalyst	Without $CuSO_4$ catalyst
s	cm^3	cm^3
s	cm^3	cm^3
s	cm^3	cm^3
s	cm^3	cm^3
s	cm^3	cm^3
s	cm^3	cm^3
s	cm^3	cm^3
s	cm^3	cm^3
s	cm^3	cm^3
s	cm^3	cm^3

Project 4: Investigating the Activity of Metals When Producing Hydrogen

You Will Need

As for the previous project plus different metal samples, e.g.:

- Magnesium
- Aluminum/aluminium
- Zinc
- Iron
- Tin
- Lead

Method

Warning

Potassium, sodium, lithium, and calcium all react violently with dilute sulfuric acid and dilute hydrochloric acid. It is dangerous to put these metals into an acid.

We want to try and eliminate as many "stray variables" as we possibly can, by ensuring that as many things are the same as possible. When selecting your metal scraps, try and look for scraps which are as near as possible the same shape and size—as we want to try and ensure that the surface area available for reaction is the same for all the metal samples. (This has an impact on the rate of reaction.)

We are going to repeat the previous project; however, this time, we are going to use different metals each time. We want to pick a selection of

scraps that are all roughly the same size and have the same surface area. We are going to take measurements with our timer and graduated gas jar, and note them in Table 3-3.

Conclusion

What did you notice? You should have seen that some metals produced hydrogen gas very quickly, while some took their time to produce gas—the metals reacted with the acid at different rates, i.e., their reactivity was different.

We can place metals into what is called a "reactivity series"—this allows us to classify how vigorously different metals will react with air, water, and acids.

Table 3-3
Measure the activity of different metals.

Time (seconds)	Magnesium	Aluminum/ aluminium	Zinc	Iron	Tin	Lead
s	cm³	cm³	cm³	cm³	cm³	cm³
s	cm³	cm³	cm³	cm³	cm³	cm³
s	cm³	cm³	cm³	cm³	cm³	cm³
s	cm³	cm³	cm³	cm³	cm³	cm³
s	cm³	cm³	cm³	cm³	cm³	cm³
s	cm³	cm³	cm³	cm³	cm³	cm³
s	cm³	cm³	cm³	cm³	cm³	cm³
s	cm³	cm³	cm³	cm³	cm³	cm³
s	cm³	cm³	cm³	cm³	cm³	cm³
s	cm³	cm³	cm³	cm³	cm³	cm³

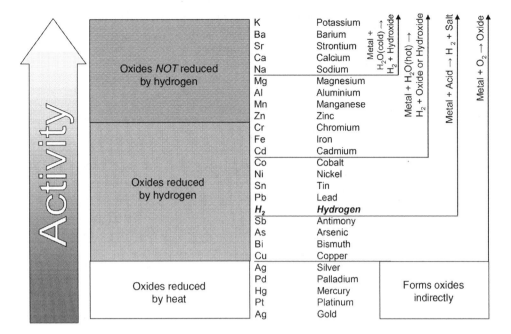

Figure 3-6   *The activity series of metals.*

Project 5: Investigating Water Electrolysis

You Will Need

- 2 × carbon rods or 2 D cell zinc carbon batteries
- 9 V PP3 cell
- Wire
- Small tub
- Baking soda
- Gloves

Method

First of all, we need two carbon rods for our electrodes. If you can get a couple of carbon rods, then great, but if not we can salvage some. We are not going to be using the D cell batteries as batteries, we only want them for their rods inside! Take a hacksaw (make sure you wear gloves for

Figure 3-7   *Baking soda adds free ions.*

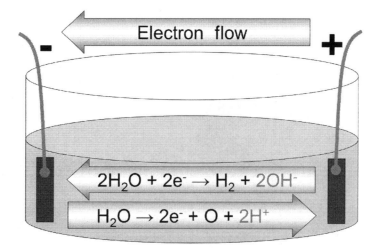

Figure 3-8 *Hydrogen at the cathode, oxygen at the anode.*

this bit) and cut the metal casing of the battery. The carbon rod is in the middle of the battery surrounded by electrolyte goo. You need to clean the rod, disposing of the spent battery casing, and the goo, sensibly.

Once you have two clean rods, connect them to either terminal of a 9 V battery using short lengths of wire. Now immerse them in a solution of water in a small tub.

What happens? . . . Very little! The reason is that ordinary water is a pretty poor conductor of electricity. We need to add some free ions to the solution.

This comes in the form of common baking soda. Add a sprinkle to the solution, and see the bubbles form!

You can capture these bubbles with a test tube, and test the respective anode and cathode to see if the bubbles are hydrogen or oxygen.

Conclusion

Free ions are necessary in a solution, for electrolysis to take place. In large electrolyzers, we might use a strong alkali-like potassium hydroxide, which is highly caustic.

So what's happening here?

Figure 3-8 gives us a guide as to the chemical reactions that are taking place at the anode and the cathode. We can see that the positively charged ions, the cations, move toward the cathode, while the negatively charged ions, the anions, move toward the anode. It is the power supply to the electrolyzer that provides the energy for this movement to take place.

We can prove that the hydrogen and oxygen are coming from water by summing the chemical reactions that are taking place at the anode and the cathode. This is shown in Figure 3-9.

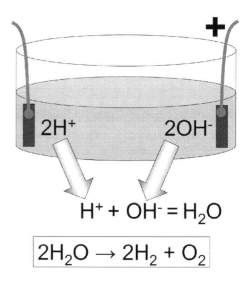

$$H^+ + OH^- = H_2O$$

$$2H_2O \rightarrow 2H_2 + O_2$$

Figure 3-9 *The complete reaction.*

Project 6: Making Fuel from Water with a Hoffman Apparatus

You Will Need

- Hoffman apparatus

Another way that we can perform electrolysis is with a piece of glassware called a Hoffman apparatus—this contains a reservoir which we fill with water and something to provide free ions. There are then two electrodes at the bottom of two large tubes, which have gas taps fitted to the top of them.

As the water is electrolyzed, the hydrogen and oxygen accumulate in the tubes.

Hoffman apparatuses were used in the past as electricity meters when supplies were in D.C. It was possible to tell the amount of power that had passed through the Hoffman apparatus by looking at the quantity of gas that had been produced.

Producing hydrogen from renewables

If we want a sustainable energy economy, then we can produce hydrogen from renewable energy using the process of electrolysis.

PURE Energy Centre

This is one of the things that the PURE Energy Centre in Shetland is hoping to demonstrate.

The Shetland Islands have a fantastic wind resource, and the folks at PURE take this wind, and turn it into hydrogen. They can then store

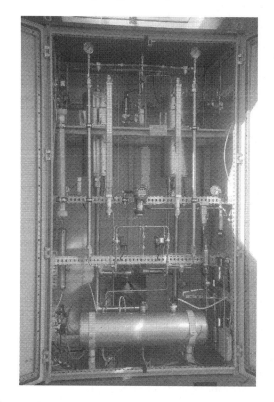

Figure 3-11 *The PURE Energy Centre electrolyzer.*

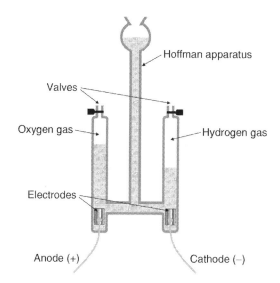

Figure 3-10 *Hoffman apparatus.*

Figure 3-12   *Two 6-kW proven wind turbines.*

this as K-type cylinders, and utilize it in a number of onsite applications, including a fuel cell car, a hydrogen barbecue, and a 5-kW PEM fuel cell.

They have a substantial electrolyzer (shown in Figure 3-11)—this takes spare electricity which is not used for the—offices from the two 6-kW proven wind turbines. As the power used to generate the hydrogen is from renewables, no climate-damaging carbon emissions are produced in the process.

HARI–The Hydrogen and Renewables Integration Project

Another project, where electrolysis is being used to turn renewable energy into hydrogen, is at HARI, based at West Beacon Farm in Leicestershire, UK.

Again, HARI employ an electrolyzer, as seen in Figure 3-13, to turn the electricity from the portfolio of renewables into hydrogen gas.

Figure 3-13   *The HARI electrolyzer. Image courtesy Dr Rupert Gammon*

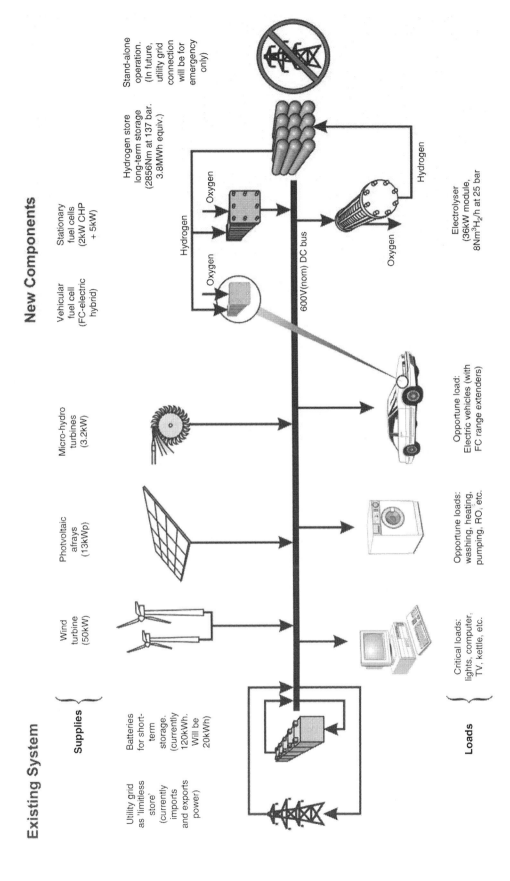

Existing System

Supplies

Utility grid as 'limitless store' (currently imports and exports power)

Batteries for short-term storage. (currently 120kWh. Will be 20kWh)

Wind turbine (50kW)

Photovoltaic arrays (13kWp)

Micro-hydro turbines (3.2kW)

New Components

Vehicular fuel cell (FC-electric hybrid)

Stationary fuel cells (2kW CHP + 5kW)

Hydrogen store long-term storage (2856Nm at 137 bar. 3.8MWh equiv.)

Stand-alone operation. (In future, utility grid connection will be for emergency only)

Oxygen

Hydrogen

Oxygen

Hydrogen

600V(nom) DC bus

Oxygen

Hydrogen

Electrolyser (36kW module, 8Nm³H₂/h at 25 bar

Loads

Critical loads: lights, computer, TV, kettle, etc.

Opportune loads: washing, heating, pumping, RO, etc.

Opportune load: Electric vehicles (with FC range extenders)

Figure 3-14 *System map of the Hydrogen and Renewables Integration Project. Image courtesy Dr Rupert Gammon.*

HARI has a diverse array of onsite renewable generating options, including wind turbines, solar cells, and micro-hydro power. The spare electricity is then converted into hydrogen, which can be used in the onsite stationary fuel cell, or in a proposed hydrogen vehicle which is currently under development.

Chapter 4

Hydrogen Storage

While the technological problems of developing fuel cell technology to a stage where it is commercially viable are being solved one by one, there are still a number of other obstacles to the implementation of a hydrogen economy.

In a future hydrogen economy, as well as looking at methods to use hydrogen, e.g fuel cells, we also need effective ways to store and transport hydrogen.

The thing is that the present sources of energy that we use have a high energy density by volume at room temperature. An enormous amount of energy can be stored in a gallon of petrol for the small size that it takes up. By contrast, the same quantity of hydrogen would take up a much larger area—by its nature, it is a gas!

While hydrogen provides very favorable figures when considering "energy per weight," the figures for "energy per volume" do not compare as favorably to traditional hydrocarbon fuels.

There are some inherent challenges in storing hydrogen, due to the fact that as a molecule H_2 is very small. Because of this, it can permeate many materials that would at first glance seem suitable materials for construction of hydrogen storage vessels.

At atmospheric pressure, the size of tank needed to store a quantity of hydrogen, compared to a hydrocarbon fuel with the same amount of energy, means that the hydrogen tank is many times larger than the conventional hydrocarbon tank, so any weight saving as a result of using hydrogen is more than negated.

As a result, there is a lot of investigation into storing hydrogen in so-called "metal hydrides." The easiest way to think of hydrides is like a "hydrogen sponge" which soaks up hydrogen, but then releases it when heated. Metal hydride storage cylinders are shown in Figure 4-21 (page 48).

Project 7: Using Hydrogen Storage Tanks with Desktop Fuel Cells

You Will Need

- For our hydrogen experiments in this book, we will be using small lightweight cylinders.

An empty storage tank is shown in Figure 4-1, and a full tank in Figure 4-2.

Figure 4-1 *Empty hydrogen storage tank.*

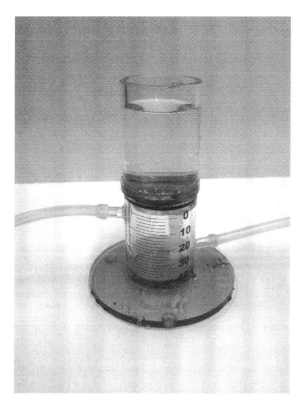

Figure 4-2 *Full hydrogen storage tank.*

Project 8: Boyle's Law and Compressed Hydrogen Storage

You Will Need

- Syringe

This is a very, very simple experiment, but it should help to firmly impress upon you Boyle's Law.

Boyle said that if a fixed quantity of gas is at a fixed temperature, its volume will be inversely proportional to its pressure.

If you take a syringe, fill it with air, and put your finger firmly over the outlet to prevent any air from escaping, you have trapped a fixed quantity of gas inside that syringe. Now, if you depress the syringe plunger, you are changing the pressure inside the syringe. You will notice that as you push further, it becomes harder to press—this is because the pressure of your fingers is acting against the volume of gas trapped inside the syringe. The volume has decreased, but the pressure has increased.

There's a really natty online simulator, which allows you to experiment with Boyle's Law from the comfort of your web browser.

Check out:

www.chem.iastate.edu/group/Greenbowe/sections/projectfolder/flashfiles/gaslaw/boyles_law_graph.html

Project 9: Exploring Charles' Law

You Will Need

- Balloon
- Freezer/fridge
- Boiling water
- Bowl
- Measuring tape

The relationship between temperature and the volume of a gas is a crucial one for us to understand if we are interested in hydrogen storage, as it allows us to do clever things to increase the volume of gas that we can store within a given space.

Charles' Law states: "At constant pressure, the volume of a given mass of an ideal gas increases or decreases by the same factor as its temperature increases or decreases."

To demonstrate this, take an ordinary balloon and tie it up so that the air cannot escape. We now have a fixed volume of gas trapped inside the balloon. Wrap the measuring tape around the balloon and measure the circumference at the widest point. Now,

put some boiling water in the bowl and immerse the balloon to allow the boiling water to heat the balloon, and thus the air inside the balloon. Then take it out of the bowl and immediately measure it around the circumference. Then take the balloon, and shut it in the fridge for several minutes. Take it out and measure it again. Then, put the balloon in the freezer and after several minutes, take it out and measure it again.

You will see that at higher temperatures, the sealed volume of gas inside the balloon expands, and the balloon increases in size, while, when the balloon is chilled, the balloon contracts. Cryogenic storage is one method proposed to store hydrogen for use in fuel cell cars. When hydrogen is stored cryogenically, if we can get it to a temperature of 20 Kelvin or $-253°$ C, the gas changes phase into a liquid. Unfortunately, there is a loss of energy in cooling the hydrogen, which equates to around 30% of the energy stored.

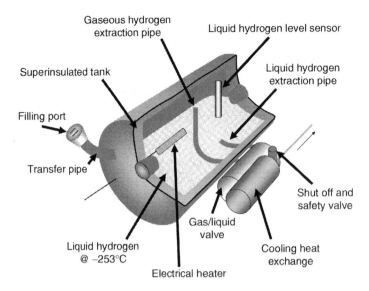

Figure 4-3   *Cryogenic hydrogen storage tank diagram.*

Labels in figure:
Gaseous hydrogen extraction pipe
Liquid hydrogen level sensor
Superinsulated tank
Liquid hydrogen extraction pipe
Filling port
Transfer pipe
Shut off and safety valve
Gas/liquid valve
Liquid hydrogen @ −253°C
Cooling heat exchange
Electrical heater

You will see in the diagram in Figure 4-3 a cryogenic storage tank of the kind we might find in cars of the future powered by hydrogen.

By cooling hydrogen, we can make it more dense, and so a fixed volume of gas at room temperature will take up a smaller space. This is useful to us when looking at how to fit lots of hydrogen into a small vehicle fuel tank!

www.

Online simulation is a good way of finding out about Charles' Law. As with Boyle's Law, there is another great online simulator at:

www.chem.iastate.edu/group/Greenbowe/sections/
projectfolder/flashfiles/gaslaw/charles_law.html

Project 10: Making Your Own "Carbon Nanotube"

You Will Need

- Drinking straws/paper art straws
- Permanent marker
- Large round stickers

One of the ways in which we can store hydrogen is to *adsorb it* onto another material, as shown in Figure 4-4. Carbon nanotubes are one example of a chemical with a high surface area.

Tools

- Hot melt glue gun (optional)

The building block for our carbon nanotube is a buckyball. The chemical formula for the buckyball is C60—this means that if we use two stickers, one either side of a straw to form a representation of a

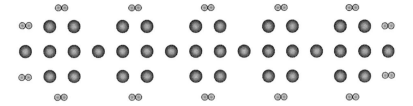

Figure 4-4 *Diagram of hydrogen adsorbed onto a surface. The lighter colored smaller circles represent the hydrogen atoms, while the larger, darker circles represent the carbon atoms.*

carbon atom, we will need 120 stickers to make our C60.

We represent carbon as being black, so add a permanent marker for coloring the stickers. We will be using straws for bonds.

Cut your straws into short lengths that are all the same. In order to produce a bond, hold the straws together in a junction, and then stick a sticker on either side of the join. Start by making a pentagon shape, with five protruding pieces of straw. Then begin to turn these additional lengths of straw into hexagons. The completed buckyball will look like Figure 4-6.

Extending our buckyball further, we can make a carbon nanotube, which is a long tubular structure, that can be used to adsorb hydrogen.

Hydrogen storage in hydrides

Metal hydrides

Metal hydrides were discovered by accident at the Philips Laboratories in Eindhoven, The Netherlands in 1969. The researchers found that a specific alloy exposed to hydrogen gas would absorb the gas like a chemical sponge. Like squeezing a sponge causes it to release the water stored within, hydrides can be "squeezed" by applying a little heat. The storage of water is therefore a reversible process. The technology of metal hydrides has found widespread

Figure 4-5 *The materials you will need to make your carbon nanotube.*

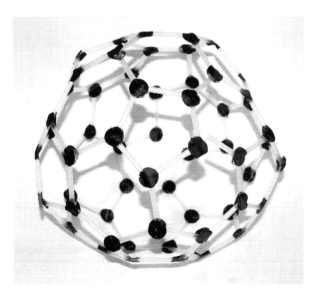

Figure 4-6 *The completed buckyball.*

application in energy storage in the common "nickel metal hydride" batteries that can be found in many of today's mobile phones and portable electronic devices, but in a hydrogen economy could these substances find other energy storage applications—in storing hydrogen? One of the advantages of metal hydrides is that they can store a very high density of hydrogen at relatively low pressures; however, this comes at the penalty of mass. Metal hydrides can be seen in Figures 4-8 and 4-9. See how the hydrogen (small circles) is absorbed in the gaps between the large circles which represent the metal atoms.

Complex hydrides

Complex hydrides are more sophisticated than metal hydrides, and incorporate additional chemicals to improve the performance of the hydride.

Complex hydrides store hydrogen by "locking it up" chemically with metals and other substances added to improve the performance of the hydride.

When hydrogen is required, the complex hydrides are heated, releasing the hydrogen.

While complex hydrides offer the benefit of having a low volume, and offer vehicle makers the opportunity to re-charge on-board, they have the disadvantage that they have a relatively high weight compared to some other solutions, require high temperatures for operation, and can suffer from issues with providing an adequate fuel flow for some applications. As a technology, they can still be considered to be in "development." An example of the structure of a complex hydride can be seen in Figures 4-10 and 4-11. In Figure 4-10 the small circles represent hydrogen gas, the large dark circles represent metal atoms, and the light colored dark circles represent additive atoms included to improve the performance of the complex hydride.

Chemical hydrides

Chemical hydrides store hydrogen in their chemical structure—releasing it when exposed to a

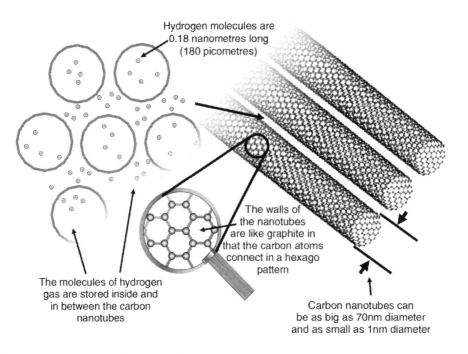

Hydrogen molecules are 0.18 nanometres long (180 picometres)

The walls of the nanotubes are like graphite in that the carbon atoms connect in a hexago pattern

The molecules of hydrogen gas are stored inside and in between the carbon nanotubes

Carbon nanotubes can be as big as 70nm diameter and as small as 1nm diameter

Figure 4-7 *Hydrogen storage in carbon nanotubes.*

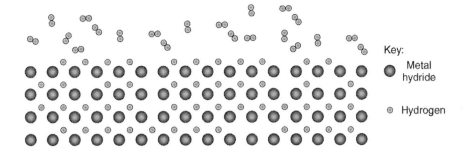

Key:

● Metal hydride

◦ Hydrogen

Figure 4-8   *2D representation of metal hydride.*

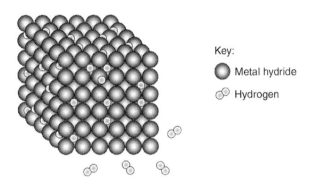

Key:

● Metal hydride

◦ Hydrogen

Figure 4-9   *3D representation of metal hydride.*

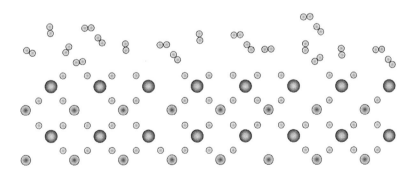

Figure 4-10   *2D representation of complex hydride.*

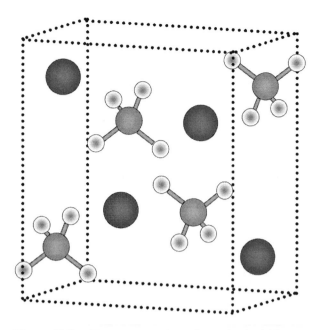

Figure 4-11 *3D representation of complex hydride.*

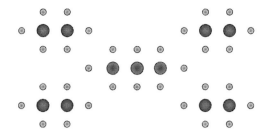

Figure 4-12 *2D representation of chemical hydride.*

catalyst, we are going to experiment with chemical hydrides later in this book using the H-Gen reactor. Chemical hydrides can be either liquids or solids. They have the potential to offer a low weight and volume solution—the fact that chemical hydrides can be distributed in liquid form opens up the opportunity of using the existing infrastructure to distribute the hydrides. The disadvantage of using chemical hydrides is that the spent hydride would need to be stored on board the vehicle, and exchanged for fresh reprocessed hydride. Reprocessing hydrides requires infrastructure, technology, and equipment. Figure 4-12 is a representation of a chemical hydride. See how other chemicals act as a "carrier" for the hydrogen.

There are such a wide range of hydride storage compounds—the array is bewildering! Looking at Figure 4-13 gives you the chemical formulas for some hydrides that have been discovered, plotting their volumetric density and density by mass. There is also a comparison to some other hydrogen storage methods.

For hydrogen to become of practical use in automotive applications, we need to improve our ability to store it. We can see how current approaches to storing hydrogen compare to the energy density of a tank of gasoline in Figure 4-14. We can also see the Department of Energy's goal for hydrogen storage, which gives us a realistic idea of how much the technology is likely to improve.

While we probably will not be able to come close to the energy density of gasoline, this is not necessarily a problem, as we can circumvent issues of energy storage by designing vehicles to be lighter, more aerodynamic, and more fuel-efficient, so that they use less energy in the first place.

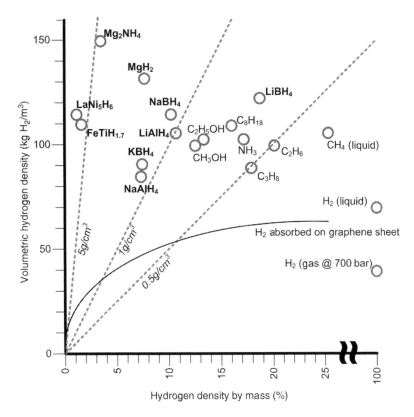

N.B. Hydrides are shown in bold

Figure 4-13   *A range of hydrides and their storage capacities.*

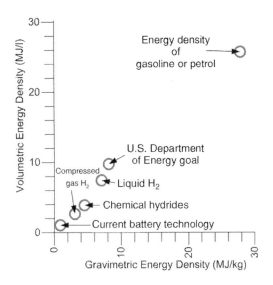

Figure 4-14   *Hydrogen storage roadmap.*

You Will Need

- H-Gen

- Sodium borohydride

A small container of sodium borohydride, like that shown in Figure 4-15, will be more than sufficient to produce much hydrogen from the H-Gen for experimentation.

We see, looking at the chemical structure of sodium borohydride in Figure 4-16, that for every sodium and boron atom, there are four atoms of hydrogen.

Method

Take the H-Gen hydrogen generator, and half fill it with water; it will only take about 15 ml. You don't

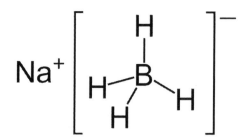

Figure 4-16 *The chemical structure of sodium borohydride.*

need anything special—ordinary tap water will do for this one.

Add about a tenth of a gram of sodium borohydride. You can use a spatula, for example, the item in Figure 4-17. At this point, you won't see a thing. Now drop the catalyst into the water. Immediately, you should see small bubbles.

The end product of this reaction is sodium borate: $NaBO_2$. We generate such small quantities that there is not a problem getting rid of it down the sink.

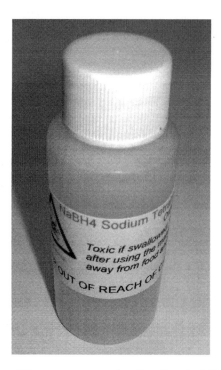

Figure 4-15 *A container of sodium borohydride.*

Figure 4-17 *Spatula used to measure the quantity of sodium borohydride.*

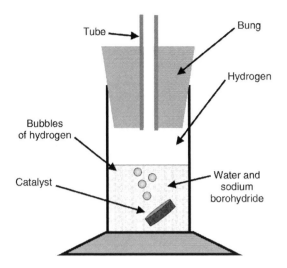

Figure 4-18 *Diagram showing the H-Gen producing hydrogen gas.*

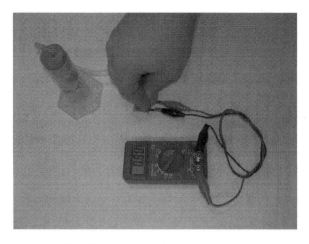

Figure 4-19 *Hydrogen produced by the H-Gen powering a mini PEM fuel cell.*

Conclusion

This simple experiment shows us that there are other ways of generating useful quantities of hydrogen besides using the electrolysis of water. The next question might be "so why don't we just use loads of sodium borohydrate to solve our energy crisis?" Well, we could, as certainly research is being done into vehicles using chemical hydrides to power them (see Figure 13-39 for more about how this would work). However, chemical hydride powered vehicles would require a different infrastructure again for refueling—as equipment would be needed to "recharge" the chemical hydrides with hydrogen gas. While chemical hydrides have the advantage of being lightweight and low-volume, the materials need to be removed from the vehicle for regeneration, and the associated technologies and infrastructure could prove expensive.

Warning

When cleaning the H-Gen, be very sure not to lose the reactors. It is these reactors that make up the bulk of the cost of the H-Gen unit. If they break, do not discard them, as they will still work.

Project 12: Changing the H-Gen Rate of Reaction

You Will Need

- H-Gen
- Timer
- Kettle
- Ice cubes

Introduction

The H-Gen produces enough hydrogen to power most fuel cells that produce up to 1.5 W of power. For some, slightly larger, fuel cells, we might need more hydrogen, or we might want to produce hydrogen at a higher rate.

In this experiment, we are going to look at the factors that affect the rate of hydrogen production in the H-Gen unit.

In order to do this experiment, we will need a means of regulating the temperature of our water. A kettle and ice cubes provide a sound way of making varying temperatures of water between 0° C (32° F) and 100° C (212° F).

Metal hydrides

Using the apparatus shown in Figure 4-20, we can experiment with hydrides, changing the temperature of the cylinders on each side, causing the pressure on each side to change. The valve in the middle can be opened or closed to allow the pressure to equalize, or build up in one side.

Figures 4-21 and 4-22 are examples of the types of hydride cylinder you could use to store

Figure 4-21 *Example 1 of metal hydride cylinders.*

hydrogen in for experimentation and work with small-scale fuel cells.

Metal hydrides are one option for storing hydrogen for transport applications; however, the technology will need to develop further if it is to provide the range that consumers want.

At the PURE Energy Centre, they have developed an electric car with hydrogen hybrid drive. A hydride cylinder is used to store hydrogen for the vehicle, and it is recharged from a supply stored in compressed hydrogen tanks, which we will look at next.

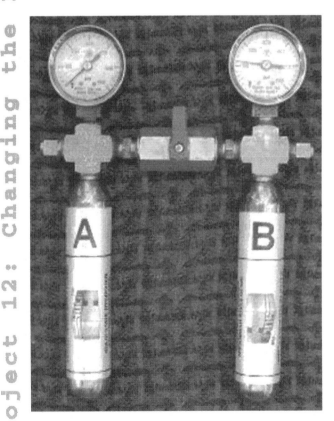

Figure 4-20 *Apparatus for experimenting with metal hydrides.*

Figure 4-22 *Example 2 of metal hydride cylinder.*

Figure 4-23 *PURE Energy Centre fuel cell car fitted with metal hydride hydrogen storage.*

Figure 4-24 *Compressed hydrogen storage at the PURE Energy Centre.*

Compressed hydrogen storage

One of the simplest ways to store hydrogen is to keep it as pressurized gas. Unfortunately, this does have the disadvantage that the gas occupies a large volume; however, it is a relatively simple approach.

Unfortunately, compressing gas requires some energy, so as fuel cell scientists, we need to be aware that the "work" required to compress our hydrogen is going to require some additional energy input, which may affect the economics of running our system.

The PURE and HARI projects both use compressed hydrogen storage, in order to store the hydrogen they produce from spare renewable energy.

At the PURE Centre, the electrolyzer works at a relatively high pressure, from which it enters the hydrogen storage cylinders.

Figure 4-25 *Hydrogen compressor at HARI project, West Beacon Farm. Image courtesy Dr Rupert Gammon.*

Figure 4-26 *Hydrogen storage at the HARI project, West Beacon Farm. Image courtesy Dr Rupert Gammon.*

At the HARI project, a slightly different approach is used, whereby the hydrogen is first compressed, using the hydrogen compressor shown in Figure 4-25 and then stored in the hydrogen cylinders shown in Figure 4-26.

By storing hydrogen in this way, the supply of energy can be matched with demand, and spare energy can be used to produce transport fuel.

Chapter 5

Platinum Fuel Cells

Introduction

We saw that Grove's original fuel cells employed platinum electrodes. You can recreate Grove's original gas battery by using a couple of lengths of platinum wire.

We are going to omit the dangerous sulfuric acid used in Grove's original experiments, replacing it instead with a mixture of water and common table salt.

From this simple experiment, you will learn the basics of fuel cell technology, which will act as a springboard to allow you to progress onto bigger and better technologies.

Project 13: Build Your Own Platinum Fuel Cell

You Will Need

- A small tumbler full of distilled water and a pinch of salt
- Voltmeter
- Battery
- Connection leads
- Length of platinum wire
- Battery

Hint

An old film canister makes a perfect little container for our platinum fuel cell.

Method

Making the platinum fuel cell

Fill a tumbler with water, and put the tiniest sprinkle of salt in it. Common household salt, such as shown in Figure 5-1, is ideal.

In order to construct the fuel cell, you will need to cut the platinum wire in half, and wrap it around a thin cylindrical object, such as the probe of a multimeter or a biro refill. This gives us a spiral that we can immerse in the water, and gives us a large surface area exposed to the water.

Two tightly wound coils of wire, such as those shown in Figure 5-2, should result.

Now, put the two coils of wire at opposite sides of the glass, and using the crocodile clips on your connection leads, clip the wire to the side of the glass, immersing it in the water. See Figure 5-3.

The crocodile clips must provide good electrical contact between the platinum wire and the clip

Figure 5-1　*Common table salt.*

Figure 5-3　*Wires positioned on either side of the beaker.*

itself, as well as supporting the wire and holding it in place (Figure 5-4).

The wire should be covered by the salt water, but the water should not touch the crocodile clips. The crocodile clips are made from metal that is "reactive" and so could form a galvanic cell (a battery), whereas the platinum will not react, and so any power being produced is as a result of the hydrogen and oxygen, not the reaction of the metals.

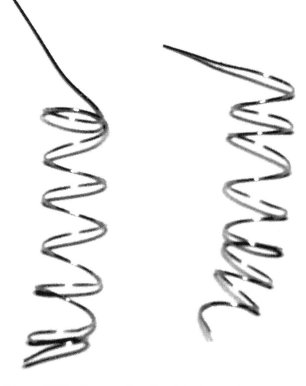

Figure 5-2　*Two small coils of platinum wire.*

Figure 5-4　*Close-up of the wires clipped snugly to the inside of the container.*

Making a fair test

We want to check that the voltage that we are producing from our platinum fuel cell really is as a result of the hydrogen and oxygen produced, not as a result of any galvanic circuit. Therefore, we must measure the voltage between the platinum electrodes immersed in the salt water solution.

Now, using the most sensitive DC voltage setting of your voltmeter, measure the potential across the two electrodes.

It should register zero.

Generating a little hydrogen (and oxygen)

We are now going to charge our platinum fuel cell, connecting the wires to a battery will cause gas to be evolved at the two electrodes. Connect three of the 2 V lead acid batteries in series. We read about this reaction—electrolysis—in Chapter 3.

A schematic diagram of the setup can be seen in Figure 5-5, and a picture of the completed setup can be seen in Figure 5-6.

Now, take a voltmeter and connect it across the two electrodes, as shown in Figure 5-7.

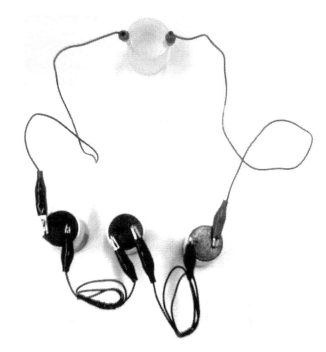

Figure 5-6 *A picture of the completed setup for electrolysis.*

As can been seen in Figure 5-8, a voltage will register on your meter. Your meter should be set to one of the most sensitive DC voltage settings.

Observe the two electrodes during and after electrolysis, and during the process of voltage being generated. What do you notice about the amount of bubbles being formed at the two electrodes? Are there more at one than at the other? Take a look at

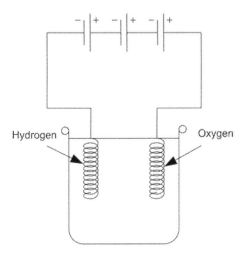

Figure 5-5 *Schematic for generating hydrogen and oxygen.*

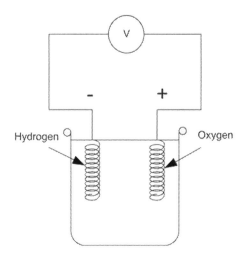

Figure 5-7 *Schematic of voltmeter connected across the platinum electrodes.*

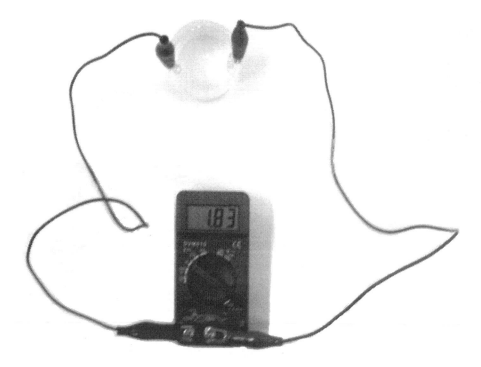

Figure 5-8 *The platinum fuel cell generating electricity.*

Figures 5-9 and 5-10, which are a close-up view of the two electrodes that were taken during electrolysis.

As we saw in the experiments on electrolysis, the positively charged ions—which we call the cations—go to the cathode, whereas the negatively charged ions, the anions, migrate toward the anode. When they reach the anode and cathode, they "give up" their charge to the electrode, completing the circuit.

Once we have formed a small amount of gas, this gas *sticks* to the electrode, a very primitive form of storage—certainly not enough storage to do anything useful on a large scale.

Once we have produced hydrogen and oxygen on the electrodes of the fuel cell, we can then "release" this imbalance of charge by creating a circuit with the voltmeter.

Then the platinum acts as a catalyst. Its catalytic action allows the hydrogen ions to break up into electrons, which flow through the electrode and the

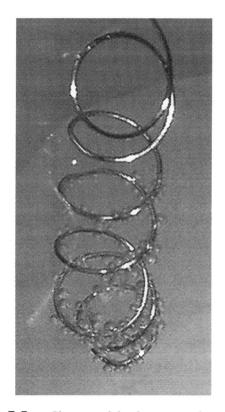

Figure 5-9a *Close-up of the the oxygen electrode.*

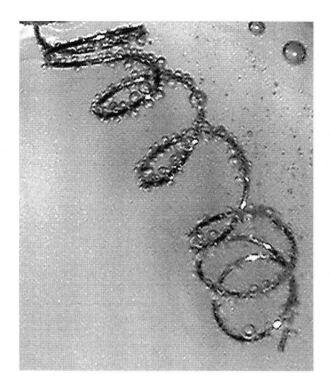

Figure 5-9b *Close-up of the hydrogen electrode.*

circuit. At the anode, the oxygen bubbles which cling to the electrode receive electrons from the electrode, and combine with the hydrogen ions that are in the water to produce water.

This process is illustrated in Figure 5-10.

Anode (+)   Cathode (-)

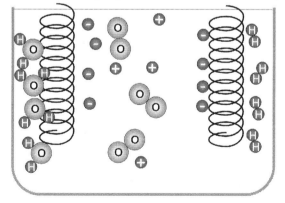

Figure 5-10 *Platinum fuel cell electrochemistry. Hydrogen is produced at the cathode and oxygen at the anode.*

Proving that the catalytic effect of platinum is required

Platinum is unique because of its catalytic properties. This is why they use it in car "catalytic converters," as well as in many industrial applications.

Unfortunately, it is also very expensive. If only we could use something other than platinum . . . Well, tough! At the moment, we can't! Try using two other wires of the same metal and, while you will see electrolysis take place, you will not see any significant voltage when connecting the meter.

If you want to take platinum fuel cells further:

There are some online movies at:

www.geocities.com/fuelcellkit/pics/FCK1.mpg

www.geocities.com/fuelcellkit/pics/FCK2.mpg

www.geocities.com/fuelcellkit/pics/FCK3.mpg

Here is another nice site to check out for platinum fuel cells:

www.scitoys.com/scitoys/scitoys/echem/fuel_cell/fuel_cell.html

Taking it further

Take a look at Grove's original gas battery in Figure 5-11. Grove used multiple cells in his battery to produce higher voltage. How could you make a platinum fuel cell with multiple cells? As a hint, you can obtain *safety* test tubes, often in chemistry kits for a younger audience. These can

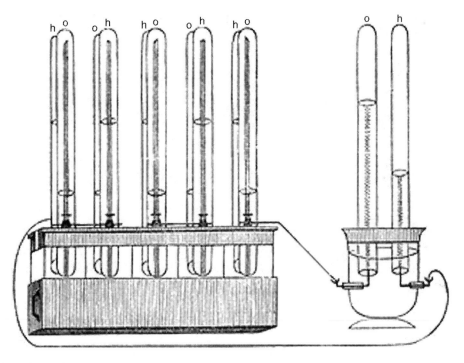

Figure 5-11 *Grove's gas battery.*

be drilled safely, unlike glass, allowing you to experiment and make multiple cells. Later in this book, we are going to look at fuel cell "stacks." Keep this in mind, and think back to Grove's gas battery when you encounter this information.

Chapter 6

Alkaline Fuel Cells

We saw in Chapter 1 how alkaline fuel cells played an important role in the early development of fuel cells, having been developed by Francis Bacon at Cambridge University. In fact, because of their inventor, you may hear them referred to as "Bacon cells." As you will know from reading this chapter, alkaline fuel cells, such as those shown in Figure 6-1, were used in early space missions, and are a very well-developed fuel-cell technology.

One of the advantages of a fuel cell in space is that you already have the hydrogen and oxygen, and the water is pretty useful for astronauts to drink too!

The chemistry of an alkaline fuel cell differs slightly from some of the other fuel cells in this book. In a PEM fuel cell, it is the protons which are exchanged through the membrane, while the electrons complete the circuit. In an alkaline fuel cell, it is "hydroxy ions" that make the journey from electrode to electrode, and the water, which is a product of the combination of hydrogen and oxygen, is produced at the anode, rather than

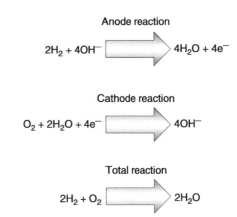

Anode reaction

$$2H_2 + 4OH^- \longrightarrow 4H_2O + 4e^-$$

Cathode reaction

$$O_2 + 2H_2O + 4e^- \longrightarrow 4OH^-$$

Total reaction

$$2H_2 + O_2 \longrightarrow 2H_2O$$

Figure 6-2 *The chemistry of alkaline fuel cells.*

the cathode—which means that the water is produced on the "hydrogen side" rather than the "oxygen side."

If we look at the pairs of chemical reactions which are taking place at each electrode (see Figure 6-2), we see that:

At the anode

Hydrogen is oxidized, producing electrons in the process.

At the cathode

Oxygen is reduced, producing "hydroxide ions," which migrate across the electrolyte of the alkaline fuel cell.

So reduction and oxidization are happening at the same time . . . Remember LEO the Lion going GER?

If you want to look at this in terms of a diagram, Figure 6-4 shows the layout of an alkaline fuel cell. Here we can see how the "exhaust" from the

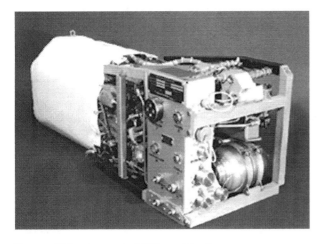

Figure 6-1 *An early AFC used in the space program. Image courtesy NASA.*

Figure 6-3   *Reduction and oxidization (remember LEO the Lion?).*

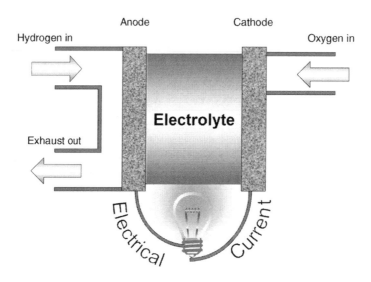

Figure 6-4   *Diagrammatic representation of an alkaline fuel cell.*

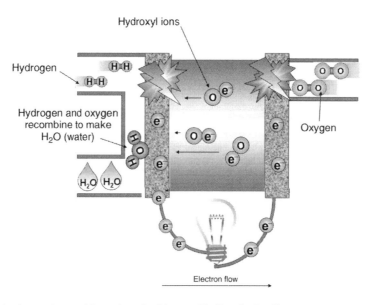

Figure 6-5   *The chemical reactions taking place inside an alkaline fuel cell.*

fuel cell comes from the anode side of the fuel cell, rather than the cathode.

If we now superimpose the chemical reactions that are taking place onto this diagram, we get Figure 6-5.

We can see that the hydrogen enters to the top right, the oxygen to the top left, and *we* can see

how electrons are flowing from the anode, to the cathode, via the circuit. Looking at the cathode, we can see how electrons and oxygen form hydroxyl ions, which flow across the electrolyte of the alkaline fuel cell, and recombine with the hydrogen protons to make H_2O, or water, which is the output from the fuel cell.

Project 14: Making Electricity with Alkaline Fuel Cells

Now, we are going to get to grips with using a *real* alkaline fuel cell. To begin with, we are just going to check that we can produce some power from our fuel cell, and as the chapter goes on, we are going to perform some simple experiments which will enable us to learn more about the performance of alkaline fuel cells.

You Will Need

- Alkaline mini fuel cell (Fuel Cell Store P/N: 530709)
- Potassium hydroxide (Fuel Cell Store P/N: 500200)
- Sodium borohydride (Fuel Cell Store P/N: 560109)
- Multimeter (Fuel Cell Store P/N: 596007)

Tools

- Small scoop
- Accurate scales
- 1 liter graduated jug

The assembled fuel cell, as delivered, should look something like Figure 6-6. You will notice that the sockets, and cutouts in the plastic on the top of the fuel cell, ensure that the two components

can only seat properly when they are correctly lined up.

The anode of the mini alkaline fuel cell is the clear outer tub, shown in Figure 6-7.

If you look inside the tub, you will see a square affixed to the bottom of the tub—this is the anode electrode, shown in Figure 6-8, and it connects to the anode terminal, which employs a 4-mm banana plug, also shown in Figure 6-8.

The red plastic piece which came supplied with the tub is shown in Figure 6-9. This is the mini fuel cell cathode assembly.

You will notice that there is a similar piece of material on the bottom of the red molding—this is the cathode electrode, as indicated in Figure 6-10. The cathode comes with a similar 4-mm banana socket, as shown also in Figure 6-10.

You can see a cross-sectional view of the fuel cell in Figure 6-11. Notice that the two electrodes are opposite each other.

Thinking about the diagram shown earlier in Figure 6-5, it should become easy to relate the components of the mini alkaline fuel cell to the diagram. We can envisage the oxygen being drawn from the center of the fuel cell. However, you will notice that there is no "hydrogen input tube." This fuel cell gets its hydrogen fuel from chemicals dissolved in the electrolyte solution.

You will remember Chapter 4, where we used sodium borohydride. Well, we are going to use it

Figure 6-6 *The assembled mini alkaline fuel cell.*

Figure 6-7 *The mini alkaline fuel cell anode.*

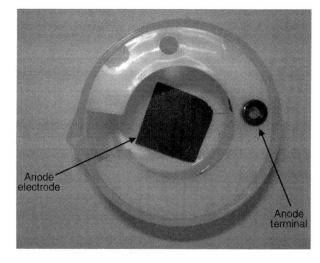

Figure 6-8 *Note the location of the anode electrode and terminal.*

Figure 6-9 *The mini alkaline fuel cell cathode assembly.*

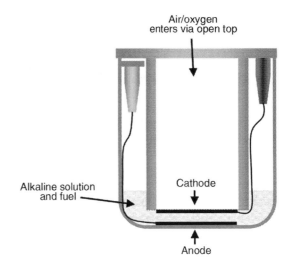

Air/oxygen enters via open top

Cathode

Alkaline solution and fuel

Anode

Figure 6-11 *Cross-sectional view of the mini alkaline fuel cell.*

Warning

If you are going to order sodium borohydride through the post, many countries and states will require that you specify "hazardous materials postage," which costs a little extra—be mindful of this when thinking about how much to order!

again as the source of fuel for our alkaline fuel cell.

You will need a small bottle of sodium borohydride, like that in Figure 6-12.

Later in this chapter, we are going to experiment with some different "hydrogen carriers," but for the moment, let's go with what works well! You will also need a small scoop for the sodium borohydride.

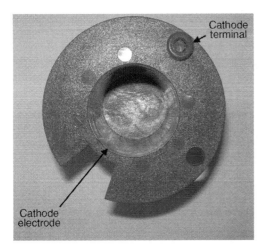

Cathode terminal

Cathode electrode

Figure 6-10 *Note the location of the cathode electrode and terminal.*

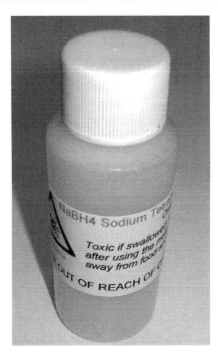

Figure 6-12 *Sodium borohydride, the source of hydrogen for our AFC.*

Figure 6-13 *Very small scoop, suitable for measuring sodium borohydride.*

Now, remember this is an "alkaline fuel cell," but "Where is the alkaline?" I hear you shout—well, it comes in the form of potassium hydroxide.

Figure 6-14 shows a bag of potassium hydroxide. If you can't get any of this stuff, then have a play around with some alternative alkalis, as shown in the later experiments—performance might be a bit impaired, but it will prove the principle.

Now, we need to prepare a 1 M solution of this. To do this, we need 56 grams of potassium hydroxide in a measuring jug—56.10564 g/mol—to which we will add water to make 1 liter.

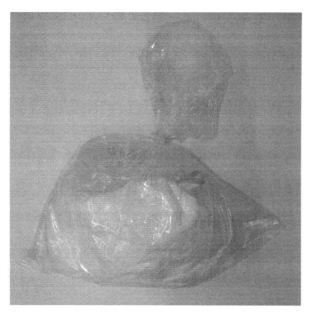

Figure 6-14 *Potassium hydroxide.*

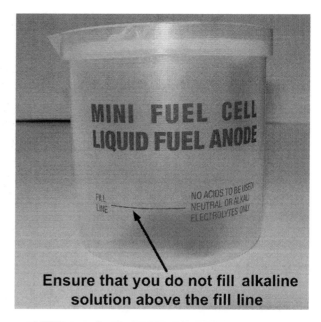

Ensure that you do not fill alkaline solution above the fill line

Figure 6-15 *Do not fill past this line!*

We need to take about 65 cm³ of this to fill our fuel cell.

Warning

If you've got 65 cm³ of the alkaline electrolyte solution, then it should all work out just fine—but whatever you do, do *not* fill past the line shown in Figure 6-15. You may need to top up a little bit to reach the line—this is fine, but do not go above it.

Hint

Don't be alarmed, don't be scared! If the liquid starts looking a bit black the first time you use the fuel cell, it is only a little bit of the carbon coming out of the cloth electrodes—don't panic, as this is normal.

Figure 6-16 *The alkaline fuel cell connected to the voltmeter.*

Add a pinch of the sodium borohydride, using the small measuring scoop. You don't need very much at all.

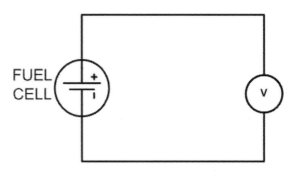

Figure 6-17 *Schematic for connecting the AFC to the voltmeter.*

Now, simply and straightforwardly, we are going to put the cathode into the anode pot, and connect the fuel cell to a voltmeter. The setup will look like Figure 6-16, and the schematic is as simple as Figure 6-17.

There really is nothing to it!

Your alkaline fuel cell should be producing power, so set your meter to its most sensitive D.C. voltage setting and look at the reading!

Project 15: Plotting the IV Characteristics of an AFC

You Will Need

- Alkaline mini fuel cell (Fuel Cell Store P/N: 530709)
- Potassium hydroxide (Fuel Cell Store P/N: 500200)
- Sodium borohydride (Fuel Cell Store P/N: 560109)
- 2 × multimeter (Fuel Cell Store P/N: 596007)
- Variable resistance

Tools

- Small scoop
- Accurate scales
- 1 liter graduated jug

In this experiment, we will be setting up the alkaline fuel cell exactly as we did in the last experiment; however, the step that will differ is how we set up the multimeters to monitor the fuel cell's performance. We will be connecting the fuel cell to a variable load—a variable resistor, and measuring the voltage across the fuel cell, and the current flowing through the load. Each time we change the variable resistance, we need to disconnect the resistor, and quickly take a reading of the resistance of the resistor, *write it down*! Don't try and remember—it never works. Then, quickly connect it back in the circuit, read the current and voltage, and *write them down too*!

In fact, I'm so adamant that writing it down is a good idea that I have even provided you with a natty little table to write them in (Table 6-1)!

We are going to be plotting a current-voltage curve for our alkaline fuel cell. To do this, we first need to obtain a series of values. So, you're going to have to connect everything up as shown in Figure 6-18. A schematic is shown in Figure 6-19.

Now, you have two options. The simple way is to use a pen and paper. I've provided you with a blank graph, so you can give it a go with plotting your points. This is Figure 6-20.

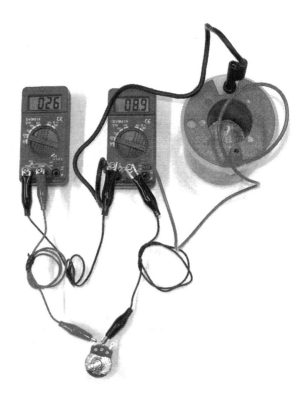

Figure 6-18 *The alkaline fuel cell measurement setup.*

Table 6-1
Record resistance, current and voltage for your alkaline fuel cell.

Resistance (R) Ω Ohms	Voltage (V) V Volts	Current (I) A Amps
Ω	V	A
Ω	V	A
Ω	V	A
Ω	V	A
Ω	V	A
Ω	V	A
Ω	V	A
Ω	V	A
Ω	V	A

Hint

You will notice that the voltage is across the *x-axis of the graph*. There is a lesson here—my science teacher, Mr Kaufman, used to patrol the room, carefully scrutinizing everyone's neatly plotted graphs. There was deadly silence. Suddenly, the silence would be broken: "Control your bottom, young man," he'd shout at some unfortunate pupil. It was nothing to do with what the student had eaten for breakfast—it was more to do with the graphs! We always put the thing that we can *control* at the bottom of the graph. In this case, we are plotting voltage and current, but we are "controlling" the voltage, by using the variable resistor. So, when plotting graphs, remember to "control your bottom."

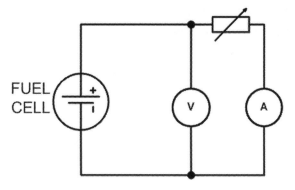

Figure 6-19 *Schematic diagram for the alkaline fuel cell measurement setup.*

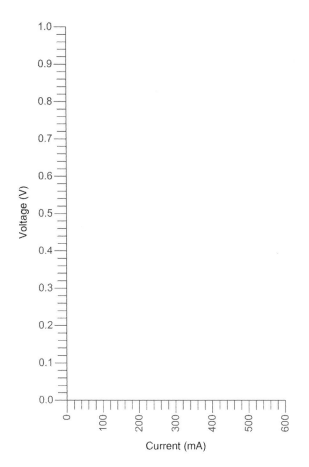

Figure 6-20 *Blank graph for plotting IV characteristic of the alkaline fuel cell.*

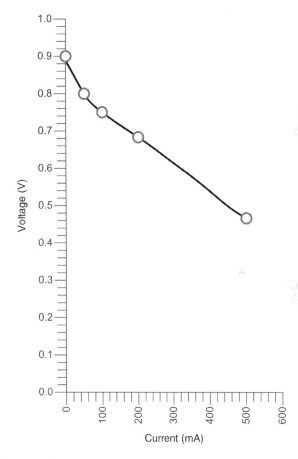

Figure 6-21 *Completed graph plotting IV characteristic of the alkaline fuel cell.*

Project 16: Plotting the Power Curve of an AFC

You Will Need

- Alkaline mini fuel cell (Fuel Cell Store P/N: 530709)
- Potassium hydroxide (Fuel Cell Store P/N: 500200)
- Sodium borohydride (Fuel Cell Store P/N: 560109)
- 2 × multimeter (Fuel Cell Store P/N: 596007)
- Variable resistance

Tools

- Small scoop
- Accurate scales
- 1 liter graduated jug

Great news! We are going to take the data from the last experiment, and process it in a different way to learn something else about the characteristics of the alkaline fuel cell.

If you haven't carried out the previous experiment, I'd suggest you do so now, as we are going to need the data.

Transfer the data about voltage and current to Table 6-2.

Once you have taken your readings and written down your series of values, you can compute the *power* column in the table. This is really simple. To find D.C. power, you simply take the voltage (in volts), and multiply it by the current (in amps), which gives you watts.

Table 6-2

Voltage, current and power for an alkaline fuel cell

Voltage (V) V Volts	Current (I) A Amps	Power (P) W Watts
V	A	W
V	A	W
V	A	W
V	A	W
V	A	W
V	A	W

We are now going to plot the power curve of our AFC. You're provided with a blank set of axes in Figure 6-22 (although bear in mind that as power changes dramatically as a result of alkaline and fuel, you may need to redraw different axes with different scales.) An example of how your power curve might look is shown in Figure 6-23.

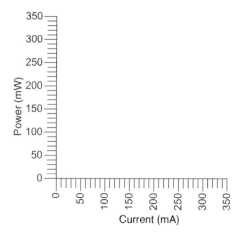

Figure 6-22 *Blank power curve graph.*

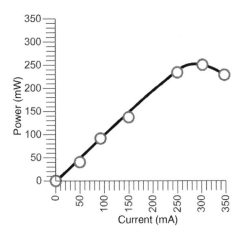

Figure 6-23   *Example power curve graph.*

Project 17: Comparing the Performance of Different Alkaline Electrolytes

You Will Need

- Alkaline mini fuel cell (Fuel Cell Store P/N: 530709)

- Potassium hydroxide (Fuel Cell Store P/N: 500200)

- Sodium hydroxide (drain cleaner)

- Sodium borohydride (Fuel Cell Store P/N: 560109)

- Multimeter (Fuel Cell Store P/N: 596007)

- Variable resistor

In this experiment, we are going to compare the performance of the alkaline fuel cell using two different alkaline solutions—potassium hydroxide (which we have been using in the other experiments), and sodium hydroxide, which is also known as drain cleaner. In order to make a fair experiment, we are going to use sodium borohydride as the fuel

for both experiments. For the first experiment, use the sodium hydroxide as the alkali. Prepare a 1 M solution—you will need to add 40 g of sodium hydroxide to a beaker, to which you add a liter of water. Use about 65 ml of this to top the fuel cell up to the fill line. Set the experiment up using the diagram in Figure 6-19. Set the resistor to a value where it registers a flow of current. Wait for the current to stabilize. Note the voltage and current produced. Now, repeat the experiment with a 1 M solution of potassium hydroxide.

You can use Table 6-3 to note your results.

Table 6-3

Comparison of AFC alkaline solutions.

	Voltage (V)	Current (A)
NAOH		
KOH		

Project 18: Comparing the Performance of Different "Hydrogen Carriers"

You Will Need

- Alkaline mini fuel cell (Fuel Cell Store P/N: 530709)
- Potassium hydroxide (Fuel Cell Store P/N: 500200)
- Sodium borohydride (Fuel Cell Store P/N: 560109)
- Methanol
- Ethanol
- A well-stocked drinks cabinet
- Multimeter (Fuel Cell Store P/N: 596007)

In this experiment, we are going to compare the performance of different liquids that we can use as hydrogen carriers in our alkaline fuel cell. We are going to use the same potassium hydroxide 1 M solution in both of our experiments. However, we are going to run one experiment using the sodium borohydride (if you did the last couple of experiments, you could use the same results). Then we are going to thoroughly clean out our fuel cell, and use a methanol solution instead; the other solution that I recommend testing is ethanol, to compare the performance with methanol. Then you can try and think of some other likely hydrogen carriers, a good suggestion being to crack open the drinks cabinet—no, not for reckless personal consumption, but to power our fuel cell. Drinks with a high alcohol content will produce power in our fuel cell.

Table 6-4
Comparison of different hydrogen carriers.

	Voltage (V)	Current (A)
CH_4O		
CH_6O		
$NaBH_4$		

Project 19: Investigating How Temperature Affects Alkaline Fuel Cell Performance

You Will Need

- Alkaline mini fuel cell (Fuel Cell Store P/N: 530709)
- Potassium hydroxide (Fuel Cell Store P/N: 500200)
- Multimeter (Fuel Cell Store P/N: 596007)
- Kettle
- Ice cubes
- Thermometer
- Methanol

In this experiment, we are going to change the temperature of our alkaline fuel cell and see how it affects the performance of the fuel cell. You need a large bowl in which you can immerse the bottom of the fuel cell. We are going to use ice and hot water to control the temperature that the fuel cell is exposed to. We are going to create a range of temperatures between 0° C (32° F) and 50° C (122° F).

Warning

You should not heat the fuel cell above 50° C (122° F) as permanent damage to your fuel cell could result.

We are going to use 10 ml of methanol solution for fuel. First of all, stick the bottle of methanol in with your ice cubes to prechill the solution. Connect the ammeter directly across the terminals of the fuel cell. Use the thermometer to take a

Table 6-5
Effect of temperature on AFC performance.

Temperature	Voltage (V)
°C/°F	V
°C/°F	V
°C/°F	V
°C/°F	V
°C/°F	V

temperature reading and note the temperature in Table 6-5. Cross out either Celsius or Fahrenheit, depending on what scale you are using. Once you have taken your readings, plot them in Figure 6-25.

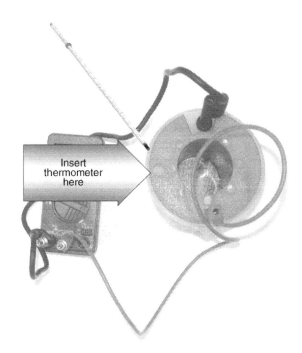

Figure 6-24   *Insert the thermometer here.*

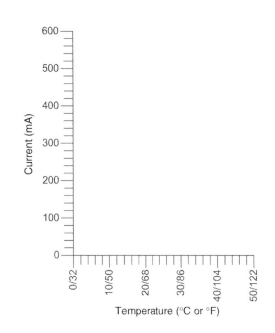

Figure 6-25   *Performance graph of AFC at different temperatures.*

You Will Need

- Vinegar
- Bicarbonate of soda
- Limewater

Tools

- Erlenmeyer flask (side tap)
- Thistle tube with bung
- Swan neck tube
- Dish

In this experiment, we are going to look at the poisoning of AFC electrolytes. One of the problems with AFC fuel cells is that they are very sensitive to impurities in the air that is supplied to them, as aqueous alkaline solutions will absorb carbon dioxide.

In this experiment, we are going to look at how an aqueous alkaline solution can be "poisoned" by carbon dioxide, but first we need to make some carbon dioxide.

We are going to make our CO_2 by reacting vinegar with bicarbonate of soda. It will fizz for a bit and produce a gas, which is our "carbon dioxide" that we are going to use to poison our electrolyte.

Our alkaline electrolyte that we are going to poison is limewater, which is saturated calcium hydroxide solution.

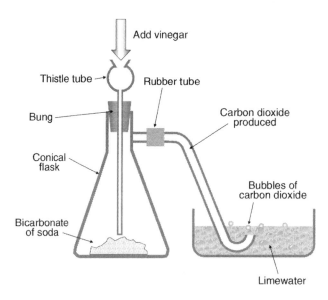

Figure 6-26   *The alkaline limewater reacts with the carbon dioxide.*

Project 21: Investigating the AFC's Need for Oxygen

You Will Need

- Alkaline mini fuel cell (Fuel Cell Store P/N: 530709)

- Potassium hydroxide (Fuel Cell Store P/N: 500200)

- Sodium borohydride (Fuel Cell Store P/N: 560109)

- Multimeter (Fuel Cell Store P/N: 596007)

In this experiment, we are going to block the passage of oxygen to the fuel cell, and see whether it can continue to function. Set up the fuel cell as in the previous experiments, with either methanol or sodium borohydride as the fuel. Then, to the inside of the red cathode, add a small amount of distilled water, just enough to cover the cathode electrode. When taking measurements, you should find that the fuel cell ceases to produce power.

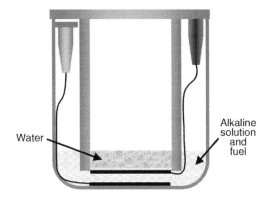

Figure 6-27   *The water blocks oxygen.*

Conclusion

This proves the need for oxygen, at the cathode, for the fuel cell's reaction to proceed.

Project 22: Investigating Different Gases at the Anode

You Will Need

- Alkaline mini fuel cell (Fuel Cell Store P/N: 530709)

- Potassium hydroxide (Fuel Cell Store P/N: 500200)

- Sodium borohydride (Fuel Cell Store P/N: 560109)

- Multimeter (Fuel Cell Store P/N: 596007)

- Selection of gases

- Bung and tube

If you have access to a range of different gases, from say a school science department, you can introduce different gases to the fuel cell's cathode, and see how the reaction proceeds. The clever design of the mini fuel cell anode means that you can take a suitably sized rubber bung, with a hole drilled through to allow a pipe to pass through, and introduce different gases to the fuel cell's cathode, as shown in Figure 6-26.

Compare how the AFC performs when using *pure* oxygen to using ordinary air. You now know how to produce carbon dioxide—so try it, and if you have access to any laboratory gases, try them too—try introducing hydrogen to the cathode.

Conclusion

You should conclude that the fuel cell needs oxygen in order to function, and that it performs better when fed with a supply of pure oxygen than when fed with air.

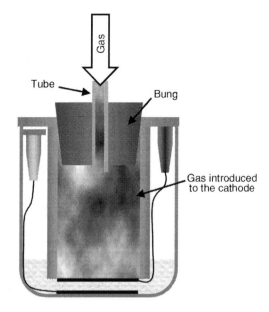

Figure 6-28 *A bung and pipe allow you to introduce different gases to the AFC's anode.*

Project 23: Investigating the Rate of Reaction (Oxygen Consumption) of an AFC

You Will Need

- Alkaline mini fuel cell (Fuel Cell Store P/N: 530709)
- Potassium hydroxide (Fuel Cell Store P/N: 500200)
- Sodium borohydride (Fuel Cell Store P/N: 560109)
- Multimeter (Fuel Cell Store P/N: 596007)
- Bung and tube
- Length of rubber hose
- Pipette
- Variable resistor
- Connection wires
- Stopwatch
- Thermometer
- Calculator

We can use the bung arrangement from the last experiment, with a short extension pipe, and a graduated tube to measure the rate of reaction happening inside the fuel cell. You should be able to make a piece of graduated pipe out of an old pipette.

The experiment is as follows. Set up the fuel cell with connection to an ammeter and a variable resistance all in series. Now, insert the bung into the cathode of the fuel cell, ensuring a tight seal. Then take the piece of pipette and insert it into one end of a piece of rubber tubing, and push the other over the length of pipe sticking out of the bung. The end of the pipette can be immersed in a small beaker of water. Figure 6-27 illustrates this.

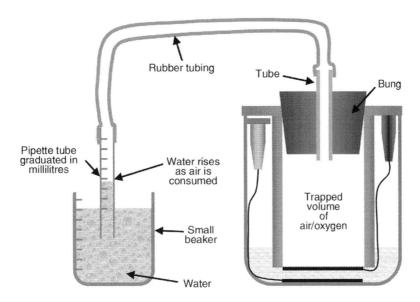

Figure 6-29 *The oxygen consumption measurement apparatus.*

As the AFC consumes oxygen as a result of the reaction, oxygen will be sucked out of the void between the cathode and the end of the pipette dipped in the water. As the AFC consumes oxygen, water will be sucked up the pipette allowing us to take measurements of how much oxygen has been consumed.

You should set the variable resistor so that it registers a reading on the ammeter—current does need to be drawn in order for oxygen to be consumed.

When you dip the pipette pipe in the water, the end of the tube will be full of air below the level of the water. As the AFC consumes oxygen, the water level in the tube will rise. When it reaches the same height as the water in the beaker, start the stopwatch.

Using the measurements on the pipette tube, we are going to time how long it takes for a volume of 1 cm³ (1 ml) of gas to be consumed. You can adjust the variable resistor, as the gas is being consumed in order to maintain a constant current flowing through the fuel cell. Once you have timed the consumption of 1 cm³ of gas, empty the end of the tube dipped in the water, and repeat the experiment again. Do the experiment three times, and enter the values in Table 6-8.

Now, work out the average of the three readings, add them together, and divide the total by three.

Figure 6-30 *Using a stopwatch, you can easily time the reaction.*

Table 6-6

Three readings for time taken to consume 1 cm³ of oxygen.

	Time (s)
Reading 1	s
Reading 2	s
Reading 3	s

We're going to do some sums now, and it makes it a lot easier if we use "standard form" for notation. If you are not familiar with standard form/scientific notation, use whatever you prefer. If you think you might need to swot up a bit, Wikipedia is always a good place to start looking: en.wikipedia.org/wiki/Scientific_notation.

Key data

Now take the thermometer, and note the ambient temperature of the room. We're going to need to convert this into Kelvins, so if you are working in degrees Fahrenheit, convert to degrees Celsius, and then once you have the figure in degrees Celsius, add 273 to convert into Kelvins.

Unless you know better (say you are on the top of a mountain), the atmospheric pressure can be assumed to be 100 kilopascals (100 kPa) or 1.0×10^5 Pa.

Now, remember we consumed 1 cm³ of oxygen—we need this figure to be in "meters cubed," so 1.0×10^{-6} m³ should do the trick!

Now, the next figure is something you are unlikely to know—it is called the *gas constant*, and it is the number that makes this equation work. For the SI units that we are using, the constant is 8.314472 m³·Pa·K^{-1}·mol^{-1}.

The Ideal Gas Law says that $PV = nRT$ where P is absolute pressure, V = volume of gas, n = number of moles, R = the gas constant and T = temperature. We can rearrange this to find out the number of moles of oxygen:

$N = PV/RT$

Now, we want to know the number of molecules of oxygen—remember Avogadro's number? 6.02×10^{23}—multiply this by the figure you obtained for the number of moles.

Next, we want to find the charge, in Coloumbs, that flowed around the circuit, so remember Q (charge, coloumbs) = A (current, amps) × t (time, seconds)—as a hint, remember that you might need to convert current readings from mA into A.

Now comes the clever bit—you should now know:

- How many molecules of oxygen were used
- How many electrons flowed around the circuit

Divide the number of electrons that flowed around the circuit by the amount of oxygen molecules concerned.

Your figure should come out as *around* 4!

Conclusion

From this experiment, we can see that four electrons are produced for every molecule of oxygen consumed. We also learn that consumption of oxygen is directly proportional to power produced.

If your thirst for knowledge about alkaline fuel cells is truly inexhaustible, you might want to check out some of these links, which make good sources of research for science fair projects, reports, essays, you name it!

—*Contd*

Apollo Energy Systems—AFC manufacturer

http://www.apolloenergysystems.com/

Astris Energy—AFC manufacturer

www.astris.ca/

Cenergie—AFC manufacturer

www.cenergie.com/

ENECO—AFC manufacturer

www.eneco.co.uk/fuelCells.html

Independent Power—AFC manufacturer

www.independentpower.biz

Intensys—AFC manufacturer

www.intensys.com

Homepage of Dr John Varcoe from the University of Surrey, who is doing very cool things with AFCs

mypages.surrey.ac.uk/chs1jv/

Chapter 7

PEM Fuel Cells

What is a PEM fuel cell? Well, it all depends on who you ask! Some will call them "plastic electrolyte membrane" fuel cells, while others will call them "proton exchange membrane" fuel cells. Whatever you call them, PEM fuel cells are *very* versatile, with the advantage that they will happily work at low temperatures.

This is what makes them useful for our experimentation purposes—we can play with them at home at room temperatures, as opposed to some of the high-temperature fuel cells in this book which require searing heat for their operation.

As with all our fuel cells, electrochemical reactions are taking place inside them, which release the energy produced as a result of the oxidation of hydrogen in the form of electricity. This is opposed to releasing it by oxidation in the form of combustion, which is horribly inefficient.

One of the disadvantages of PEM fuel cells is that they are easily poisoned by carbon monoxide, and metal ions. This makes them very sensitive to hydrogen that is not pure—and with fuel cells, damage can be expensive to repair!

How do PEM fuel cells work?

Take a look at Figure 7-1, and we'll slowly work through the operation of a PEM fuel cell. We can see the cell represented schematically in this diagram.

You can see the hydrogen entering on the anode side of the fuel cell, the oxygen entering on the cathode side of the fuel cell, and the membrane in the middle. The membrane is coated with catalyst on both sides.

We are going to break the process down into a series of steps. In reality, this is a continuous

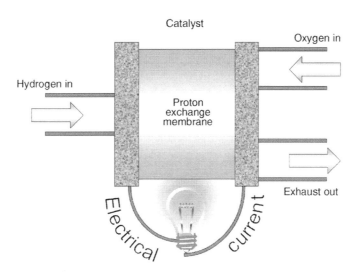

Figure 7-1 *The proton exchange.*

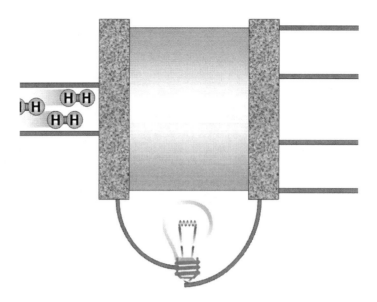

Figure 7-2 *Hydrogen enters the fuel cell.*

process which is taking place; however, looking at it in stages, rather than all at once, helps us to understand it.

We can see, in Figure 7-2, the hydrogen fuel entering the fuel cell.

When it reaches the proton exchange membrane, the catalyst splits it into protons and electrons. The protons begin to cross the membrane, while the electrons make their way around the circuit (see Figure 7-3).

As this is going on, oxygen is also entering the fuel cell on the cathode side.

As this process is happening, the protons are migrating across the solid electrolyte PEM membrane, while the electrons are journeying around the circuit (Figure 7-5). They all meet up at the cathode.

When they reach the cathode, the catalyst coating helps the oxygen, protons, and electrons to combine to form water, as shown in Figure 7-6.

Now, let's look at the components of the cell, and get to know the PEM fuel cell by taking one apart, and building a small fuel cell kit.

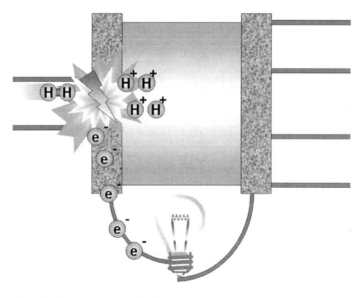

Figure 7-3 *The hydrogen is split into protons and electrons.*

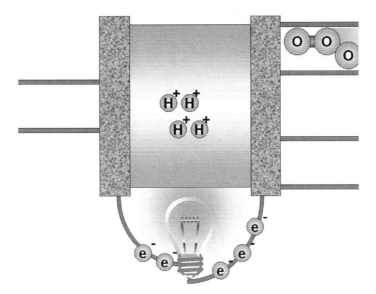

Figure 7-4   *Oxygen enters the fuel cell.*

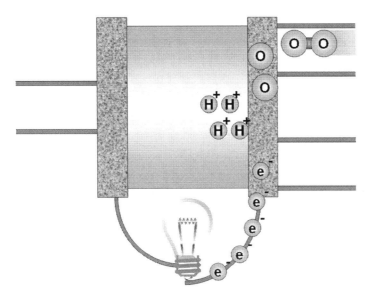

Figure 7-5   *The electrons make their way around the circuit.*

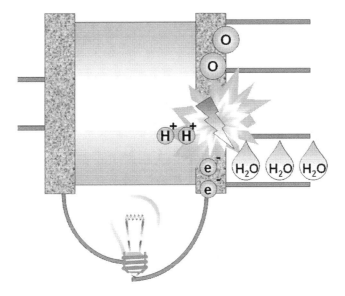

Figure 7-6   *The electrons, protons, and oxygen combine to form water.*

Let's start with the membrane

The membrane is the particularly clever piece of our PEM fuel cell—so clever, in fact, that I wonder why they don't call it a mem-brain!

In our PEM fuel cell, it is the membrane that does all the hard work, allowing certain reactions to take place, while preventing others.

To understand exactly how it works, let's take a closer look at what happens inside the membrane.

Take a look at Figure 7-7.

OK, starting in the bottom left of Figure 7-7, we can see what looks like a sandwich. On either side, we have our electrodes—the anode and the cathode, and in the middle, we have a couple of slices of gas diffusion membrane, and then, the best bit of all, our MEA!

Now, move to the magnifying glass at the bottom of the image:

You see the gray line running down the middle—well, that is our "Nafion™" membrane—think of it like a cookie with chocolate chips on both sides. The cookie dough is the special plastic our fuel cell is made from. The gray blobs either side, that look like chocolate chips, are the gas diffusion layers which are bonded to either side of the Nafion™ membrane. Where the "chocolate chips" meet the cookie dough, we have our platinum catalyst. This can be seen more closely in the two magnifying glasses at the top of Figure 7-7—you will see that one magnifying glass concentrates on the anode, while one examines the cathode.

Now, our reactions take place at the point where the cookie dough (Nafion™), chocolate chips (GDL), platinum catalyst (the small white dots between the chocolate chips and cookie dough—let's call these "sprinkles"), and gas meet. The hydrogen gas comes in at one side, permeates

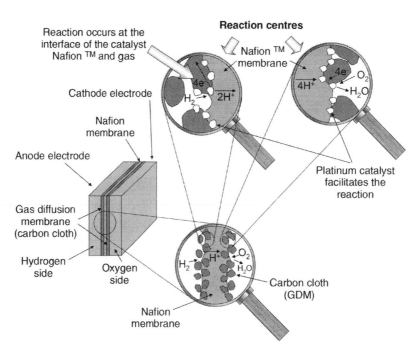

Figure 7-7 *What goes on in the middle of the PEM membrane!*

through the GDL, and the sprinkles (platinum) catalyze a reaction (and this is where the cookie [analogy] crumbles), where the hydrogen passes through the cookie (Nafion™), the electrons pass through the chocolate chips (GDL), where they are reunited at the other side of the cookie. Now, remember we said there were two sides to this cookie? (Yes … even Millies make mistakes), there are more sprinkles (platinum catalyst) on the other side, so, as our electrons are conducted through the chocolate chips to meet the cookie dough (which protons have just passed through), the sprinkles catalyze another reaction between the protons, the electrons and … introducing … *oxygen* from the other side of the cookie. They all get together to produce water!

Now, clearly we are not using cookies, and chocolate chips and sprinkles—the materials used in fuel cell construction are *much* more sophisticated (unless you evaluate them by taste—*don't try it kids*).

We're going to look at the materials used in a little more depth!

Chocolate chips

Well, although it looks like chocolate chips from the side, in reality, the woven carbon cloth of the GDL is more like a closely woven lattice of chocolate sauce. It needs to have holes to allow the hydrogen to penetrate through to the cookie dough (Nafion™ membrane); however, it still needs to provide sufficient surface area and support to the sprinkles underneath to allow the reactions to take place.

Sprinkles

We may use slightly different materials for the sprinkles on each side of the fuel cell. While hydrogen is relatively easy to split, the oxygen is a little more of a challenge. Platinum is a good material for this application, however, with advances in material science, we may find more efficient and cheaper catalysts to replace the sprinkles on both sides of the membrane.

All you wanted to ever know about sulfonated tetrafluorethylene copolymers . . .
. . . and probably quite a lot you didn't!

Our PEM fuel cells wouldn't exist if it weren't for rapid advances in polymer technology that allowed the creation of the "solid electrolyte".

Nafion™ was discovered in the late 1960s by a man called Walther Grot at DuPont™. Nafion™ has enabled the construction of the PEM fuel cell, due to its unique chemical properties.

There are a couple of things that fuel cell fans *really* like about Nafion™:

Mechanically it is very stable—it's not going to crumble when you look away, it's flexible and won't crack—it's a little flimsy, but treat it with respect and it will be reliable.

Chemically it is tolerant of a wide range of chemicals and does not react or degrade readily (apart from with alkali metals).

Thermally it is tolerant of the temperature experiences in PEM fuel cells.

It is stable to changes in temperature.

In Figure 7-8, we can see the Nafion™ unit: the top right is the Teflon backbone of the Nafion™ (the same material you find in your nonstick frying pans), while the bottom right shows the sulfonic acid group that allows for the conduction of protons.

It's hard to pin down the exact nature of Nafion™ structure, although one suggestion is the cluster model shown in Figure 7-9: Look at the sulfonic acid groups around the outside.

The SO_3 groups (or sulfonic acid) groups in the Nafion™ can *jump* from one side of the membrane to the other. This allows them to transport protons

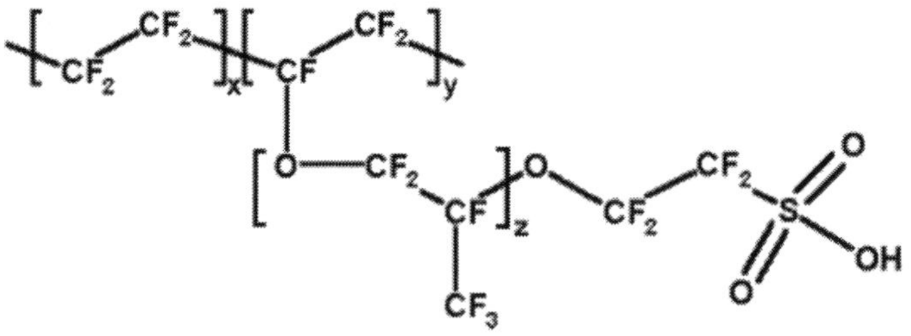

Figure 7-8 *Nafion™ chemical structure.*

from one side of our PEM fuel cell to the other—a crucial facet of the PEM fuel cell's operation.

What's also *really* clever about these little sulfonic acid groups is that they only allow the passage of cations—the anions and electrons can just get lost if they think they're crossing this sulfonated tetrafluorethylene copolymer membrane! However, as we know, the electrons might have to take a slightly more *circuitous* route (OK, so it's a bad pun!), but on their journey, they do useful work for us.

So the sulfonic acid groups act as gatekeepers in our fuel cell, only letting those with *proton passports* across the border, while the electrons have to take the long way home.

I often lay awake wondering where we would be without sulfonic acid groups …

Flow fields

In some of our smaller fuel cells, we can make do without a fully blown flow field, and get away with a piece of carbon cloth or paper to do the same job—just spread the hydrogen over the surface of the MEA, while allowing electrons to be transferred from the GDM to the electrodes.

However, once we start looking at more sophisticated fuel cells, carbon cloth and paper just doesn't cut the mustard—we need to look at more sophisticated methods of spreading the hydrogen thinly and evenly over the surface of the MEA, on the anode, and ensuring that sufficient amounts of oxygen come into contact with the MEA on the cathode side.

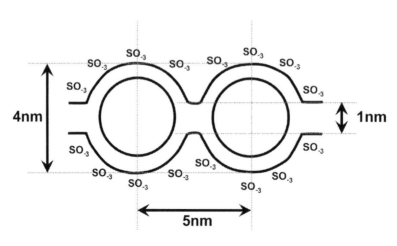

Figure 7-9 *Nafion™ cluster model.*

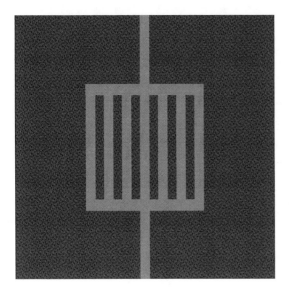

Figure 7-10   *Straight flow field.*

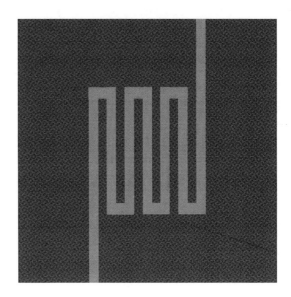

Figure 7-12   *Serpentine flow field.*

The type of flow field we select will depend on the flow characteristics that we desire. There is ongoing research into how the design of the flow field can affect the efficiency of the fuel cell, with scientists working to reconcile two competing factors—the need to spread hydrogen over the surface of the MEA, while still maintaining good electrical contact with the surface of the MEA.

There are a few different patterns that we can look at, in Figures 7-10 to 7-13.

The simplest of them all is the straight flow field, which has a place for gas to be input, and place for excess gas to escape, and a straight grid of lines between the two sides.

Getting slightly more advanced, in Figure 7-11, the interdigitated flow field has interlocking fingers of hydrogen supply and return lines. However, these are not connected directly, so the hydrogen must make a path to the MEA before spare hydrogen can be returned.

Figure 7-11   *Interdigitated flow field.*

Figure 7-13   *Spiral flow field.*

The serpentine flow field is like a long undulating snake—the hydrogen has to make a path forward and backward, forward and backward, before it reaches the return. Hopefully, by this time, much of the hydrogen has had its chance to cross the MEA's membrane.

Finally, another variation is the spiral flow field—it does exactly what it says—spirals in, then spirals out.

Of course, manufacturers are developing new variations of these flow fields every day. You might even like to turn to the project entitled "Making Your Own MEAs" on page 126, and look at how you can advance the state of the art in fuel cell technology.

Project 24: Disassembling a Fuel Cell

You Will Need

- A dismantleable fuel cell, e.g., Fuel Cell Store P/N: 80044

Tools

- Supplied Allen key and spanner

By taking apart a fuel cell, we can learn a lot about its construction, which will help us to relate some of the theory, here on the page, to the practical aspects of constructing a fuel cell. If you have any thoughts of scratch building your own fuel cell, it is a good idea to start with a commercially available dismantleable fuel cell, to get to grips with the construction.

However, be warned—although such fuel cells are designed for being taken apart and reassembled, they can still be damaged if you do not take sufficient care.

Figure 7-14 shows us a dismantleable fuel cell.

The fuel cell kit will come with a spanner and Allen key, that fits the fixings used to aid assembly and disassembly. You can see in Figure 7-15 that it is relatively easy to remove the fixings.

When all four nuts and bolts have been removed, ensure that you keep them somewhere safe, and have a flat, clean surface to work on, as in Figure 7-16.

Figure 7-14 *The complete dismantleable fuel cell.*

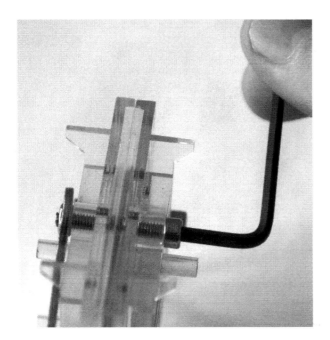

Figure 7-15 *The supplied Allen key and spanner can be used to disassemble the fuel cell.*

Lay the parts out so that you can see what is happening as you take the fuel cell apart.

When you've got all the fixings away, you can carefully remove the plastic end-plates, as shown in Figure 7-17—this will expose the electrode assembly in the centre of the fuel cell.

Lay out the end-plates and electrode assembly, as in Figure 7-18.

The electrode assembly is what is shown in Figure 7-19—it comprises of two electrodes, the gas diffusion membrane, and the membrane

electrode assembly. You now have to decide whether you want to take this component apart—while educational, you run the risk of damaging the fuel cell at this point if you are not careful.

Carefully remove one electrode and then the next, exposing the MEA. You will be left with three components, as shown in Figures 7-20 and 7-21. Ensure that you do not touch the MEA, or the surfaces that come into contact with it.

If you wish, you can use a scalpel to carefully deconstruct the MEA—it will render it useless though. Figure 7-22 illustrates the components of the MEA well. With one either end, you can see the plastic gaskets that support the MEA, with the carbon GDMs either side of the central MEA.

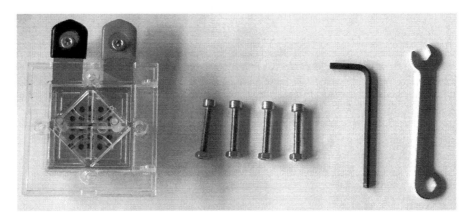

Figure 7-16 *The fuel cell with fixings removed.*

Figure 7-17   *Removing the end plate.*

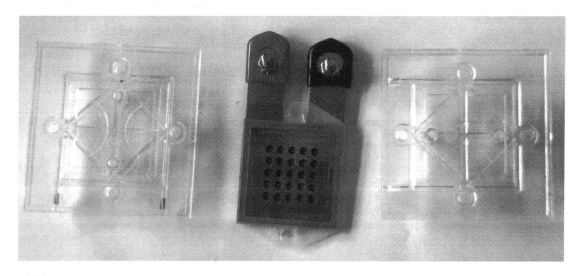

Figure 7-18   *The end plates and electrode assembly.*

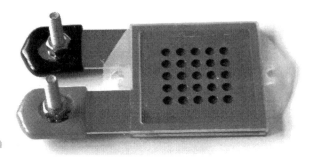

Figure 7-19   *The electrode assembly.*

Figure 7-20   *One electrode removed.*

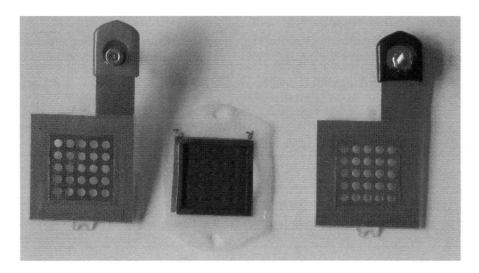

Figure 7-21 *Two electrodes removed.*

Figure 7-22 *The MEA deconstructed.*

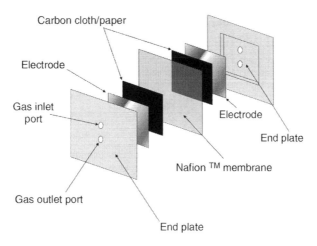

Figure 7-23 *Diagram showing the components of your dismantleable fuel cell.*

Relate the deconstructed MEA to the earlier diagrams in the chapter, compare the schematic models of the fuel cells earlier in this chapter, and relate it to your practical experience of taking the fuel cell apart.

Figure 7-23 may be of help, when considering how the different layers of the PEM fuel cell go together—providing an exploded diagram.

Project 25: Changing MEA Platinum Loading

You Will Need

- Dismantleable fuel cell—extension kit (Fuel Cell Store P/N: 80044)

The performance of a PEM fuel cell will depend to some extent on the amount of platinum catalyst sandwiched between the Nafion™ membrane and the GDL. The more platinum, the more sites there are for reactions to take place, and so the output of the fuel cell should be higher.

One way that you can experiment with this is with the dismantleable fuel cell. The extension kit comes with a couple of extra MEAs, with different platinum loadings to the standard one fitted in the fuel cell. You can substitute these MEAs, reassemble the fuel cell, and compare how the cell performs under different conditions.

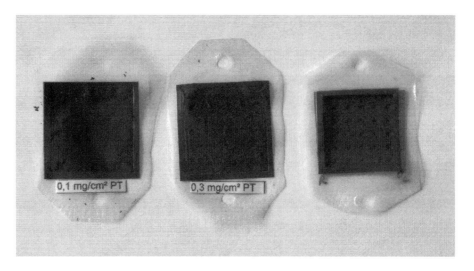

Figure 7-24   *Different membrane electrode assemblies.*

Project 26: Experimenting with Oxygen and Air

You Will Need

- Dismantleable fuel cell with oxygen and air end-plates (Fuel Cell Store P/N: 80044)

You will realize that oxygen is only a component of the air that we breathe. In fact, our atmosphere is comprised of 78% nitrogen and around 20.95% oxygen.

It should therefore follow that a fuel cell running on pure oxygen will perform in a superior manner to one running just on air.

We can experiment with this proposition, by using a dismantleable fuel cell. Remember, with a good experiment, we want to control as many factors as possible, to reduce the chance of error arising from other sources. In practice, this means that we want to keep as much of our fuel cell the same as possible—keeping the same MEA, GDM and electrodes, but changing the oxygen to air or vice versa.

When we are running our fuel cell on air, we do it at atmospheric pressure. We want to maximize the amount of air reaching our fuel cell, as it is not being forced under pressure through a pipe. So when running our fuel cell on air, we change the

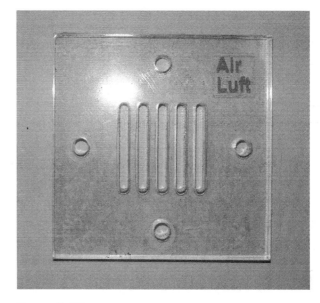

Figure 7-25 *Air fuel cell end-plate.*

end-plate with pipe connections, for one with an open grate that allows easy passage of air.

When we feed oxygen into a fuel cell, we need to do it under a very small pressure, so we can use an end-plate like Figure 7-26, which has a pair of barbs to feed and return the oxygen to our fuel cell.

Set your fuel cell up with both air and oxygen end-plates, and take measurements to compare the performance of the cells using oxygen and air.

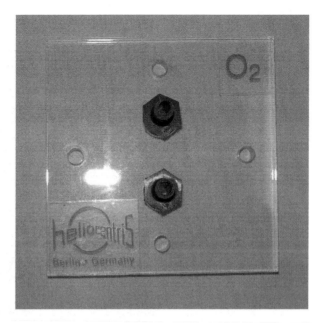

Figure 7-26 *Oxygen fuel cell end-plate.*

Project 27: Building Your Own Mini Fuel Cell

Consumer electronics devices are becoming increasingly miniaturized. You only have to look at the suitcase mobile phones of the 1980s with bricks for handsets, and compare them to today's phones, which can outsmall a bar of soap for evidence of this trend.

One of the limiting factors in miniaturizing electronic devices has been the power technology. Early mobile phones were necessarily large—not only because the electronics inside were big, but also because the batteries needed to power them were commensurately massive.

Improvements in battery technology, and the improving energy efficiency of electronic devices is leading to smaller and smaller powerpacks; however, there is a complex tradeoff with size for operating life. Smaller batteries mean that you can talk for less time on your phone, listen to less tunes before having to recharge, or watch less movies on your portable media player before it is back to the docking station.

Fuel cells potentially offer all kinds of exciting possibilities for powering portable electronic devices.

You Will Need

- Mini fuel cell (Fuel Cell Store P/N: 531907)

Tools

- Pair of hands
- Cotton gloves (optional)

The mini fuel cell is a really clever construction. The manufacturers have used the gasket material to keep all the parts of the fuel cell together, with simple folding required for assembly.

Figure 7-27 *The mini fuel cell laid out.*

You can see, the mini fuel cell laid out in all its glory in Figure 7-27. We can see by the perforated metal electrodes looking at the ends. You will see that the metal is "dished in" slightly—this is to ensure that it makes good contact with the GDM when the fuel cell is folded. You can see the two tiny GDLs moving into the next two squares. This provides electrical contact between the electrodes and the MEA, while allowing the hydrogen (and don't forget the oxygen) gas to diffuse toward the MEA .... And finally, our central square is the membrane electrode assembly—the little patch where all the exciting reactions take place!

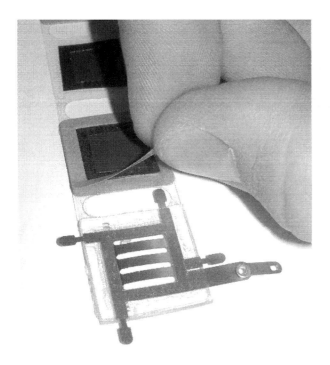

Figure 7-28 *Peeling the backing paper from the adhesive gasket.*

Start at one end, and work your way along the mini fuel cell—the squares of brown adhesive paper are removable protective coverings for an adhesive layer. You need to peel these off one by one, and fold the fuel cell over in sequence, making sure you don't touch any of the sensitive bits of the fuel cell, for risk of poisoning the catalyst.

Cotton gloves will protect the elements of the fuel cell that can be damaged, but if you've got clumsy fingers like me, you're going to struggle to peel the adhesive backings from the squares, and you might end up getting in a mess with cotton gloves stuck to the sticky bits of the fuel cell.

As you remove the backing paper from the adhesive layer adjacent to one electrode, fold that section of the fuel cell over, and ensure that the GDM makes good contact with the electrode.

Right! This step is absolutely crucial. Two things you need to remember: do not touch the MEA under *any* circumstances. Cotton gloves are a godsend in this respect, although when peeling off the fiddly adhesive backings, they can also be a nightmare—use your judgment and caution. Once you've teased off the adhesive backing from your GDM, you are going to need to line up the flimsy MEA with the square of GDM. Ensure that you get them to sit directly over each other for optimum performance. Figure 7-30 shows the fiddly job of lining them up.

OK, it's sharp nails at the ready for the last time! Tease the final corner of that last adhesive square, and fold the fuel cell over for the last time, leaving you with a neatly packaged little square mini fuel cell.

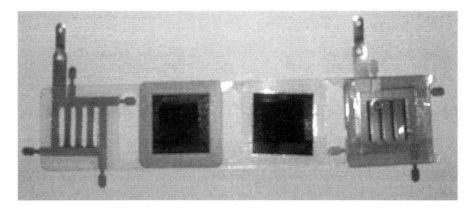

Figure 7-29 *The fuel cell once folded.*

Figure 7-30 *Lining up the MEA.*

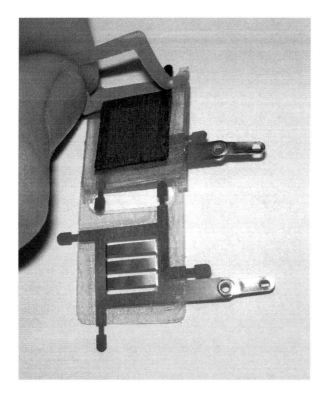

Figure 7-31 *Removing the final adhesive backing.*

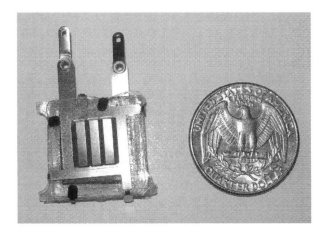

Figure 7-32 *Mini fuel cell next to a quarter for scale.*

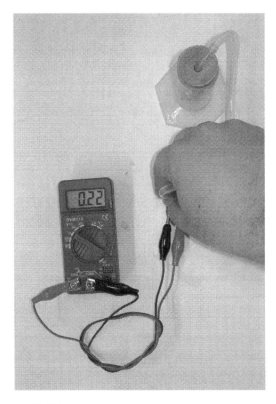

Figure 7-33 *Use a multimeter to check power is being produced.*

Once you've stuck the fuel cell together with the adhesive squares, you need to ensure that proper contact between all the layers is made. You will see that all around the edge of the mini fuel cell, there are small metal extensions to the electrode. These are folded over the opposing electrode, nice and tightly, to clamp the fuel cell assembly together. Don't worry, they won't short out against the opposite electrode—the plastic end-plates will stop this from happening!

And just to remind you of the wonders of the mini fuel cell's miniaturization, check out Figure 7-32, with the fuel cell next to a quarter for a sense of scale. This is what your mini fuel cell should look like when finished. Note the tabs folded over.

Now, you should be proud of your workmanship. Go and test it out! Grab yourself a multimeter, and a source of hydrogen. Hold the end of the hydrogen tube close to the small hole in the plastic end-plate of the mini fuel cell, and watch it crank out the volts.

Yes! It really does work. To see a video of the mini fuel cell powering a "spinner" powered by the hydrogen from the H-Gen generator, check out my video on YouTube at:

www.youtube.com/watch?v=JsWfvUktNuo

Chapter 8

Direct Methanol Fuel Cells

One of the problems with hydrogen is that it is a little problematic to store. This is a concept that we will explore in the later chapter on hydrogen storage. There is another type of fuel cell, which is able to utilize methanol as a carrier of hydrogen.

There are a number of advantages to this approach: methanol has a very high energy density and it can be transported easily using existing infrastructure for moving liquid fuels.

One of the problems with hydrogen is that it requires a massive change of infrastructure. While this is not impossible, it does necessitate a change. One of the advantages of the methanol fuel cell is that the methanol can be stored and transported using the conventional infrastructure that is used to move and transport other liquid fuels such as petrol. Because the methanol can utilize the existing infrastructure, many believe that methanol will be a good transitionary fuel in the move over to a hydrogen economy.

Attention

Direct methanol fuel cells require a solution of methanol in water, if the concentration of methanol is too high you will damage the membrane in the fuel cell.

Manufacturers recommend that you work with concentrations that do not exceed 3%. In practice, this is often a little overcautious, and concentrations around 4% can work successfully. However, note that concentrations much higher than this will certainly damage your fuel cell, and concentrations over 3% are not recommended for prolonged use.

First aid guidelines

If you get methanol in contact with your skin, you will need to wash it off with copious amounts of water, as methanol is toxic.

Protect your eyes from splashes from methanol by wearing safety goggles at all times. If you do inadvertently splash even the tiniest amount of methanol into your eyes, you will need to thoroughly wash them using clean water or an eyewash station, and contact a doctor immediately or your hospital's ER department.

Be careful not to inhale methanol fumes. If you do, open windows, and walk outside and get some

Warning

Warning! Methanol is TOXIC and highly flammable.

Keep methanol away from all sources of ignition. Only operate the system in an area that is sufficiently ventilated to ensure that you are not exposed to high concentrations of methanol fumes.

fresh air. If symptoms persist, consult your doctor or your hospital's ER department.

Methanol should not be ingested. In the event that you do accidentally ingest some methanol, drink lots and lots of clear water and consult a doctor immediately or your hospital's ER department.

If you feel nauseous, consult your doctor. Tell the doctor you have been working with methanol, and show him the material's data sheet.

Direct methanol fuel cell chemistry

Direct methanol fuel cells directly *reform* the methanol, extracting the hydrogen, and producing carbon dioxide with the *spare carbon* in the methanol. We can see methanol fuel cell chemistry in Figure 8-1.

Looking to the diagram in Figure 8-2, we can see how the methanol fuel cell is a membrane fuel cell, just like the PEM fuel cell we looked at in the previous chapter. Functionally, they are very similar.

Overlay the chemical reactions that are taking place (see Figure 8-3), and we can see water and methanol entering together at the top left, the protons being stripped of their electrons with carbon dioxide, and water the resulting product. The protons pass through the membrane, the same as in the PEM fuel cell, and the electrons pass round the circuit. On the other side of the diagram, things are the same as we are used to seeing with the PEM fuel cell.

We can see that the carbon dioxide is produced at the anode of the fuel cell.

Hint

Get a good measuring cylinder, like the one in Figure 8-4, as you will need to prepare methanol solutions of different concentrations.

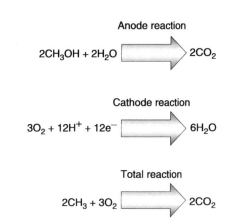

Anode reaction

$$2CH_3OH + 2H_2O \longrightarrow 2CO_2$$

Cathode reaction

$$3O_2 + 12H^+ + 12e^- \longrightarrow 6H_2O$$

Total reaction

$$2CH_3 + 3O_2 \longrightarrow 2CO_2$$

Figure 8-1 *Direct methanol fuel cell chemistry.*

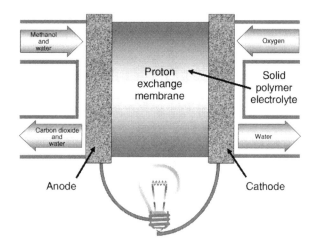

Figure 8-2 *Direct methanol fuel cell diagram.*

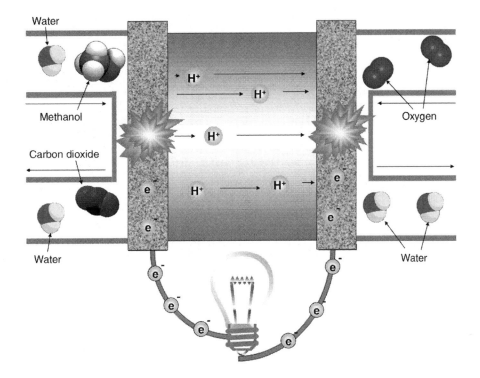

Figure 8-3 *What goes on inside a DMFC?*

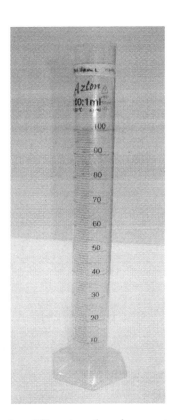

Figure 8-4 *A cylinder suitable for preparing different methanol concentrations.*

You Will Need

- DMFC fuel cell and load (e.g., DMFC and fan Fuel Cell Store P/N: 550307)
- Prepared methanol solution—3% methanol in distilled water (Fuel Cell Store P/N: 500100)
- Squeezy bottle

If your fuel cell has been stored in a dry atmosphere, you may find its initial performance a little disappointing. This is because the membrane needs to be hydrated. Try using a little distilled water first of all to dampen the membrane, and then fill the cell with some methanol solution.

The methanol fuel cell has two holes at the top of the assembly. These holes allow you to fill the fuel cell with a prepared methanol solution. You can pour methanol into one hole, and the other hole will allow air to escape.

The correct way to fill the fuel cell is to use a squeezy bottle with a narrow spout. Hold the fuel cell so that the two holes are angled, insert the squeezy bottle into one hole, and squeeze until there is no air left in the DMFC chamber.

Now pour some 3% methanol solution into the cell.

The cell will take a little time to start up. It is not unusual for it to be several minutes before the cell starts producing any power. Once the cell starts producing power, it will happily run for a few hours.

Once the cells stops producing power, then the methanol has been exhausted. The cell must now be drained and refilled to start producing power again.

Figure 8-5 *Closeup of filling the methanol fuel cell.*

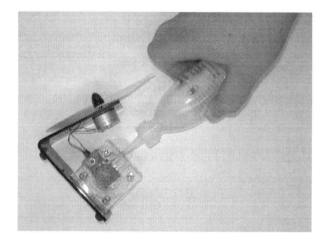

Figure 8-6 *Filling the methanol fuel cell.*

Project 29: Decommissioning and Recommissioning a Methanol Fuel Cell

You Will Need

- Direct methanol fuel cell
- Distilled water
- 1% sulfuric acid
- 3% methanol solution

Introduction

If a methanol fuel cell is not to be used for some time, it should be "decommissioned" and then "recommissioned" when you are ready to use it again. This will ensure long life of the cell.

Method

Decommissioning

Flush the cell out with some distilled water that does not contain methanol. Now, using a small piece of tape, seal the two holes. This will prevent the cell from drying out.

Recommissioning

If the cell is to be used again, the tape should be removed. First, try the cell with some 3% methanol solution, recharge it a couple of times, and leave it for several minutes.

The cell will degrade if it is left without being used. If you find that it will not work even after recharging with 3% methanol, then you will need to recommission the cell.

Warning

Sulfuric acid is very nasty stuff and precautions should be taken to ensure that you do not get it on your hands, in your eyes, or ingest it.

Prepare some 1% sulfuric acid solution and charge the fuel cell as you would do with methanol. Leave it for a couple of days. Then empty the acid and recharge with a 3% methanol solution. The cell should work again.

You Will Need

- 2 nice big Band-Aids®
- Type 316 stainless steel wire cloth (Fuel Cell Store P/N: 590663)
- 5L DMFC MEA 5cm² (Fuel Cell Store P/N: 590310)
- Cotton gloves (Fuel Cell Store P/N: 591463)
- 3% methanol solution (Fuel Cell Store P/N: 500100)
- Multimeter (e.g., Fuel Cell Store P/N: 596107)

Tools

- Scalpel
- Scissors

In this experiment, we are going to build a very simple, rough and ready, direct methanol fuel cell with the help of a box of Band-Aids®. I originally published this project in *MAKE:zine*, and many fuel cell experimenters have found it a simple, easy introduction to fuel cell technology. I really do not think there is a simpler way to scratch-build a direct methanol fuel cell, and get to grips with the technology.

Before we get stuck into the nitty-gritty of construction, I must share with you, esteemed reader, how this project came into being. I had just read on the net, the quote from Lawrence Burns, the GM vice president for research and development, who said, "The fuel cell offers a cure, not a Band-Aid." Shortly after, while going to the bathroom, my eyes clapped on a box of Band-Aids,

and an idea began to form . . . The result is this project, which is truly a Band-Aid!

Once you've gathered all the parts in the list, you should have an array of bits that resemble Figure 8-7.

You need to ensure that you have suitably large Band-Aids.

Think Rite Aid or Walgreens. These are just your bog-standard Band-Aids. Don't get ones that are impregnated with aloe vera or any special ointment—it will simply gunk up the works. Just get plain old Band-Aids. You want one with a nice big area for the "wound."

Remember, from the point you start peeling the adhesive film from the Band-Aid, you need to keep the surfaces that come into contact with the MEA scrupulously clean. At all times when working with the MEA, you will need to wear cotton gloves as shown in Figure 8-10. This is because the muck on your hands will greatly inhibit the function of the MEA.

We are going to make our electrodes from stainless steel screen. If you want to get really techie

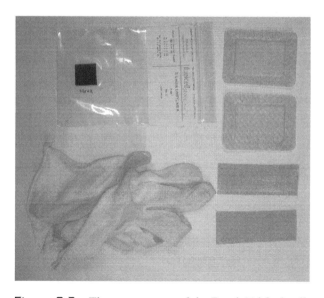

Figure 8-7 *The components of the Band-Aid fuel cell.*

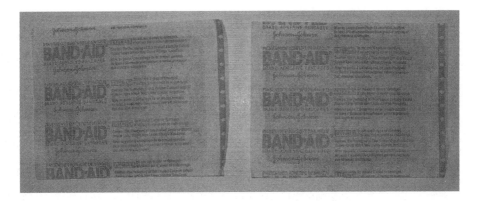

Figure 8-8   *A pair of suitably sized Band-Aids.*

and heavy about stainless steel fly screen, the material that I used was top-notch stuff. The stainless steel was 318 grade, i.e., very high quality! The screen was 72 wires per inch in both directions, with each wire being 0.0037 inch. To show you the sort of material we are referring to, look at Figure 8-11.

MEA—the membrane electrode assembly

MEA stands for "membrane electrode assembly"—this is the bit that "does the works."

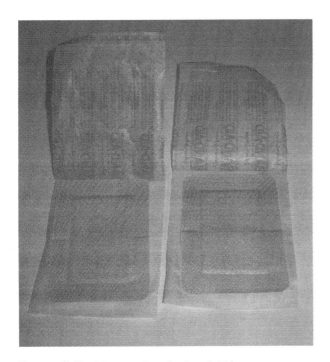

Figure 8-9   *Unwrapping the Band-Aids.*

You can make your own, but they are rather difficult to assemble—much simpler is to buy one that is already made from www.fuelcellstore.com. This is what you get for your money: a piece of Nafion™— this posh plastic is technically known as a sulfonated tetrafluorethylene copolymer . . . Let's look a little more closely at the plastic bag in Figure 8-12.

The large DMFC tells you that the fuel cell is a direct methanol fuel cell—this is a variant of a proton exchange membrane fuel cell. What it essentially means is that the MEA reforms the methanol to carbon dioxide and hydrogen without any external action required. It really is all clever stuff! Looking at the panel on the right-hand side of the label, we can see that the "active area" of

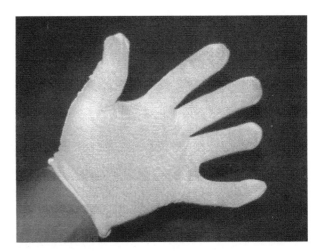

Figure 8-10   *Cotton gloves ensure the MEA is not damaged.*

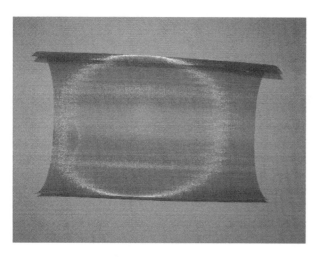

Figure 8-11 *Stainless steel fly screen.*

the fuel cell—that is to say, the bit with the catalyst—is 2.3 cm × 2.3 cm, about an inch square for those using Imperial. The Nafion membrane itself is significantly larger—5.5 × 5.5 cm square, or about 2 inches square.

We can see looking at the details below that the anode is coated with a mixture of platinum and ruthenium, 4.0 milligram per cm^3 to be exact, while the cathode is loaded with 2.0 milligram of platinum. The membrane has a built-in gas diffusion layer of carbon cloth.

Making the Band-Aid fuel cell

Making the Band-Aid fuel cell is a real walk in the park. All you need to do is be sure that you protect the MEA and handle it delicately, and try and ensure all surfaces remain clean.

First of all, you need to cut a rectangle of wire mesh. This should be a little narrower than the smallest dimension of the wound pad of your Band-Aid, and longer than the longest dimension of your Band-Aid. Cut another identical strip.

Now, remove the protective covering from one Band-Aid, as in Figure 8-13, and place the first piece of wire mesh so that it sits just inside the wound pad of your Band-Aid. The other end will protrude past the end of your Band-Aid. It should resemble Figure 8-14.

Now, remove the MEA from the bag carefully using cotton gloves (see Figure 8-15).

You need to trim the MEA so that it is slightly bigger than the wound pad of your Band-Aid, but so that there is still a generous border of Band-Aid sticky to hold the assembly together. The MEA *must* overlap the wound pad (see Figure 8-16).

Figure 8-12 *Details of the MEA we are using for the Band-Aid fuel cell project.*

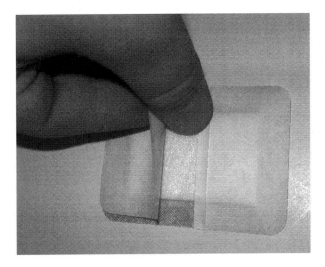

Figure 8-13 *When removing the cover of the Band-Aid, do not touch the wound pad.*

The next shot (Figure 8-17) displays the MEA checked for size against a Band-Aid. Note that "anode" is marked on the MEA—you need to make a note of on which side "anode" appears the correct way round on; make a note of this and put a "+" on the Band-Aid facing the MEA. You might want to use an overhead projector pen to do this.

Next, lay the MEA on the wire mesh (see Figure 8-18), noting orientation. Ensure that the overlap of the MEA sticks to the border of Band-Aid adhesive—you need to isolate each side of the MEA from each other using the Band-Aid adhesive as a barrier.

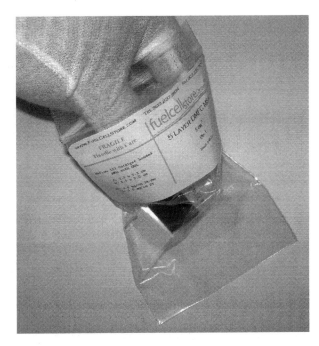

Figure 8-15 *Getting the MEA out of the bag.*

Now take the other strip of wire mesh that you cut, and lay it up on the MEA so that the spare length of gauze protrudes from the opposite end.

I found at this point that it was a good idea to bend the wire gauze back on the Band-Aid, corresponding to the side of the MEA the gauze was in contact with. This helped me to remember which piece of gauze was what electrode.

Now, use a scalpel to remove a 1-cm square of the *pink* bit of Band-Aid from the side that is in

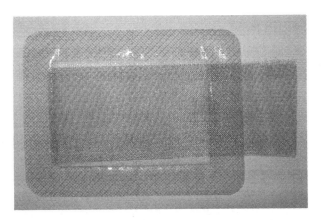

Figure 8-14 *Laying the wire mesh on the first Band-Aid.*

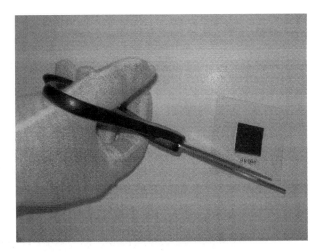

Figure 8-16 *Trimming the MEA to size.*

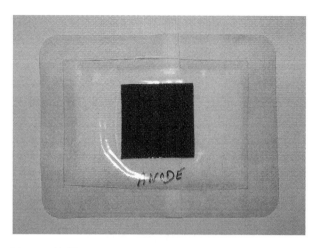

Figure 8-17 *Checking the MEA for size.*

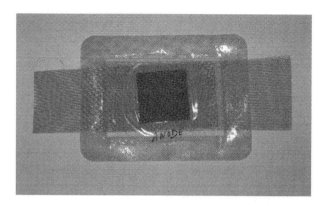

Figure 8-19 *Laying the final layer of wire mesh on top of the MEA.*

contact with the anode side of the MEA (see Figures 8-20 and 8-21).

Testing the Band-Aid fuel cell

To test the fuel cell, you will need some 3% methanol solution. Again, you can get this from fuel cell stores, unless you know a bit about chemistry or are friendly with a college lab technician.

To fuel the cell, you will need to introduce methanol solution to the patch of Band-Aid you removed on one side. The wound patch will act as a reservoir for the fluid. Keep the other side of the

fuel cell dry, and oxygen will permeate through the breathable Band-Aid.

Connect the electrodes protruding from either end of the Band-Aid to a multimeter, but do not expect big shakes first time round! DMFC fuel cells take time to "break in."

I found that the voltage that my cell produced was improved by applying a little light pressure to the area either side of the methanol top-up point. This is because the wire gauze makes better contact with the MEA.

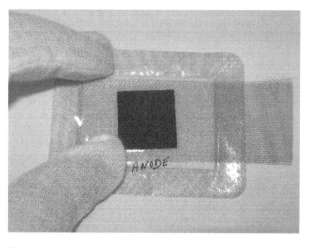

Figure 8-18 *Laying the MEA on the wire mesh.*

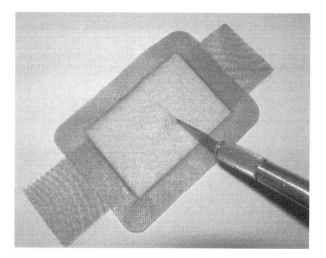

Figure 8-20 *Using a scalpel to remove a square from the top layer of the Band-Aid (initial cut).*

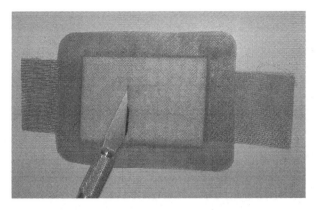

Figure 8-21   *Using a scalpel to remove a square from the top layer of the Band-Aid (completing the cuts).*

The future of DMFCs

You'll see in our Band-Aid fuel cell that we had a separate anode and cathode—and it's important that the methanol only touches the anode. Well, a company caused CMR has developed an innovative DMFC fuel cell (see Figure 8-24), which takes away some of the complexity of separating anodes and cathodes in a DMFC stack.

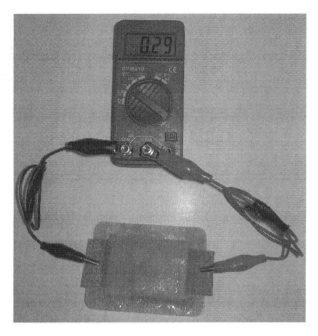

Figure 8-23   *The Band-Aid fuel cell connected to a meter.*

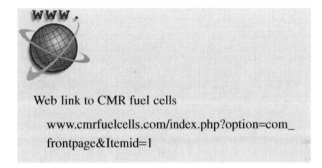

Web link to CMR fuel cells

www.cmrfuelcells.com/index.php?option=com_frontpage&Itemid=1

The CMR DMFC fuel cell is different, in using selective catalysts on the anode and cathode side. Their DMFC is fed a bubbled solution of air and methanol solution. However, because the methanol will only react with the selective catalyst on the anode, and the oxygen will only react with the selective catalyst on the cathode, no physical separation is necessary—this is the sort of innovation that is going to reduce the complexity of fuel cells, bring down their price, and help them to reach the consumer electronics market more quickly.

We can carry out many of the same experiments in the PEM fuel cell chapter, using simple DMFC fuel cells which we can connect in a circuit, such as that shown in Figure 8-25.

Figure 8-22   *The completed Band-Aid fuel cell.*

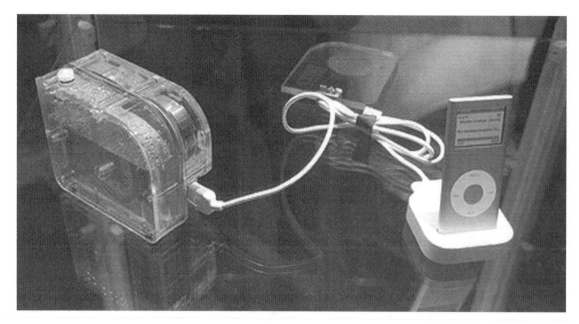

Figure 8-24   *CMR's innovative DMFC fuel cell.*

Figure 8-25   *DMFC fuel cell suitable for experimentation.*

Chapter 9

Microbial Fuel Cells

In this book, we have looked mainly at fuel cells which rely on physical, chemical reactions taking place. In this chapter, we are going to look at a radically different type of fuel cell, which operates in a radically different way—the microbial fuel cell.

The feedstock for microbial fuel cells is a "bioconvertible substrate." This is jargon for a substance that can be converted biologically into energy for the fuel cell. This can comprise a range of things—sugar (glucose), or biomass feedstocks specifically grown for the purpose, such as algae.

Living creatures take food, and metabolize it to provide them with energy. They are oxidizing energy rich, which (reading ahead to LEO the Lion goes GER) implies that they are electron rich.

The process of digestion is a very complex one, involving lots of reactions which are catalyzed by enzymes. This process happens at relatively low temperatures. In the reference section, the paper by Bennetto describes this process as "cold combustion."

History of the microbial fuel cell

In 1912, Professor M.C. Potter, at the University of Durham, managed to extract electricity from *E. coli* bacteria—this work was duly published in the proceedings of the Royal Society, but then was ignored. It took until 1931 for the idea to be revisited. Barnet Cohen took another look at the idea and managed to produce a number of cells, which, when connected together, produced a potential difference of 35 volts, albeit at a *very* low current.

Where is the technology going?

Bacteria take in food and produce waste products. Some bacteria take a "fuel" such as glucose, and digest the fuel to produce two waste products: carbon dioxide and water, in the presence of oxygen. If you remove the oxygen, the bacteria tend to produce carbon dioxide and protons and neutrons.

Medical science has now progressed to the point where it can prolong life, or improve the quality of people's lives by implanting electrical devices in the body to help assist the function of impaired organs. Pacemakers and cochlear implants are examples of medical devices that are implanted "subdermally," that is to say, "below the skin." These are all candidates that could potentially be fueled by microbial fuel cells in the future.

Scientists at Tohoku University have developed fuel cells which will use blood as a fuel. This paves the way for medical implants that can be powered by the patient, negating the need for future operations to replace batteries.

NASA estimate that over the duration of a journey to Mars, taken to be six years, a crew of six astronauts could produce up to six tons of solid organic waste, with a lot of that being human excrement. Clearly in the depths of space, you can't just FedEx it back home to mission control, so what do you do?

Well, microbial fuel cells could help to digest this waste to turn it into useful electricity. This electricity can be used to power the mission, and through the process of biological digestion, the output product is easier to process.

However, microbial fuel cells are also starting to find practical applications in the here and now. Fosters, Australian brewers of "the amber nectar," have installed a microbial fuel cell in their brewery, which takes the waste products from the beer-making process—sugars, yeasts, and alcohols—and digests them, producing clean electricity in the process. The output is cleaner waste-water, without so many of the biologically active components—and free electricity into the bargain!

Project 31: Building Your Own Microbial Fuel Cell

You Will Need

- 2 × P-traps (available from a plumber's merchant or builder's store)
- Short length of waste pipe (4 in/10 cm) to fit P-traps
- Plastic wrap
- Agar jelly
- Salt
- Carbon cloth (available from Fuel Cell Store P/N: 590642)
- Copper wire
- Aquarium aerator pump
- Length of aquarium tube
- "Bubbler" block
- Epoxy resin adhesive
- Bung or plug for P-trap top pipe or cooking oil

Tools

- Scissors
- Junior hacksaw (for cutting plastic)
- Stove
- Pan

The best way to learn about how microbial fuel cells work is to construct a simple version. This makes an ideal science fair project, and is great because it traverses the disciplines of biology and physics.

Making the MFC chambers

For our microbial fuel cell, we need to create two separate distinct chambers, interlinked by a pipe called a "salt bridge." There is a wealth of information from different experimenters (shown in the links section), who have constructed MFCs using found items, bottles, plastic tubs, pots, etc. to form their two chambers. This is a perfectly acceptable solution; however, you may find that you encounter difficulty when joining the salt bridge to the sides of the other tubs—often lots of epoxy and or silicone is required, and it is hard to get a good leak-free join.

If you've got a little bit of money in your experimenter's budget, nip to your local plumber's supply outlet, and ask for a couple of P-traps—these are the gadgets that fit underneath your washroom sink, and prevent nasty smells coming up from the drain. These provide an easy method of joining the salt bridge to them, as they come with a coupling, with a rubber gasket to ensure tight, leak-free joins.

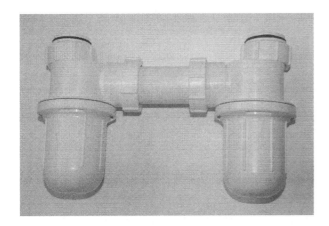

Figure 9-1 *Trial assembly of the P-traps and salt bridge pipe.*

To get the best out of a P-trap, you will see if you open them up that there is a plastic pipe extending into the cavity inside the trap. To make as much room as possible for your experiment, take a multitool such as a Dremel, and remove this plastic protrusion, giving you an empty chamber to work with.

You can see in Figure 9-1 what a neat solution a pair of P-traps and short length of waste pipe make for constructing MFCs.

Making the salt bridge

Our salt bridge separates the reduction and oxidization sides of our microbial fuel cell, while connecting them electrochemically. The salt bridge will allow a flow of ions between the two chambers of our MFC.

It consists of a simple tube, filled with a mixture of salt and agar for gelification. If you're wondering where to get agar from, it is sometimes used as a vegetarian substitute for gelatin, or you might be using it in the science lab to grow cultures on.

To make the salt bridge, you'll need to take the short length of pipe, and wrap one end firmly with several layers of plastic wrap to seal it. Ensure that you make a good seal, and use several layers of the wrap pulled taught over the end. You could even

run a bit of tape around as well to secure. Place the tube sealed end down.

Now, put the pan of water on the stove. You are going to need about 125 ml of agar solution to fill your salt bridge. However, for practicality you may need to make more than this in order to heat evenly in your pan. As a ready reckoner, for every liter of water, you're going to need about 100 g of agar. Add the agar once the water is boiling, and then throw in some salt.

When you're done, and before the solution starts to gelify and set, pour the mixture into the 4 in. 100 mm length of tube you just covered with plastic wrap, and allow it to solidify and set.

When you have allowed the jelly to set completely, loosen off the clamps on either end of the P-traps, squeeze the ends of the pipe into the fixing, and tighten the clamps up. The rubber gasket trapped between the big nut and the P-trap will ensure a good seal with no leaks.

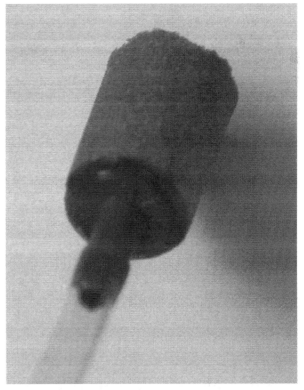

Figure 9-2 *Aquarium aerator "air brick."*

Making the electrodes

Take two sheets of carbon cloth, and fix a piece of wire to them. An epoxy resin glue is good for this. You want to ensure that there is conductor to carbon cloth contact, and that the glue doesn't interfere with this interface. The glue is applied *over* the joint between the carbon cloth and wire, not between it.

Test the connection between the wire and electrode using a multimeter set to the resistance setting. The reading should be very low.

Obtaining the bacteria for your fuel cell

You are going to need to find some bacteria for your fuel cell. You can experiment with different cultures of bacteria, but it is essential that when you are collecting them, you preserve your own health by wearing gloves. You could try collecting sediment from a river.

We are going to try to exclude oxygen from our anode, and in our cathode, we are going to try and create an oxygen-rich environment, using an aquarium aerator—a small pump that blows bubbles into an aquarium to keep the little fishies happy!

You are going to need to exclude oxygen from the anode chamber. There are a couple of ways of doing this. You can get some sort of bung or cap, or another alternative is to float some oil on top of the bioconvertable mixture, which will prevent oxygen from reaching it. Water and oil are not miscible, and the oil will just float on top of the water-based solution.

The pump forces air down a small piece of aquarium tube; however, we also need an additional device on the end of the tube called an "air brick." The air brick (Figure 9-2) is a small piece of porous ceramic that takes the feed of air coming from the aerator, and breaks it up into a plethora of little bubbles. The effect of an aerator brick can clearly be seen by looking at the pipe submerged

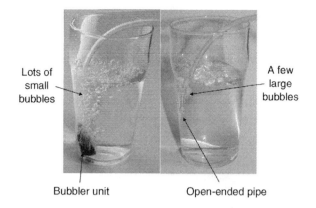

Figure 9-3 *With and without aerator bricks.*

in a glass (see Figure 9-3). When the plain pipe has air blown through it, there are a small number of large bubbles, however, when the aerator brick is added, there are a large number of small bubbles.

This is important, because it is the surface area of the bubbles in contact with the water that we are trying to maximize to increase the rate of the reaction.

Powering up your MFC

Remember that you marked one chamber "A" for anode and another "C" for cathode. Well, this is important now as you are going to need to add the right mixtures to the right chamber.

To the anode:

You are going to add the inoculum—this is our bacterial mix.

To the cathode:

We are going to add some electrolyte—salt water solution.

Connect a multimeter to the terminals of your MFC—a really easy way to do this is to solder some banana plugs onto the ends of your multimeter leads, so you can connect them directly to a meter.

Looking a little closer, we can refer to Figure 9-5 for a diagram of what is going on. Figure 9-5 is a simplification of our P-traps and assembly, and it schematically represents the MFC. In our anode

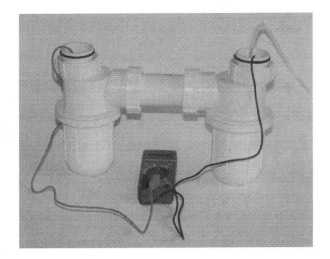

Figure 9-4   *The completed microbial fuel cell.*

chamber, we have our bacterium, isolated from air and our anode electrode.

Inside, the bacteria are digesting food and, in the process, giving up electrons to the anode terminal. The anode is "gaining" electrodes; therefore, a process of reduction is happening at this electrode.

The bacterium that we use in this fuel cell (Figure 9-6) must be a mediatorless bacterium, as we did not add any additional chemicals to "mediate" the transfer of electrons from the bacterium to the electrode.

Remember LEO the Lion goes GER? Loss of Electrons is Oxidation, and Gain Equals Reduction. So by losing electrons to the electrode, the process of oxidation is occurring. The electrons then make their way around the circuit to the cathode.

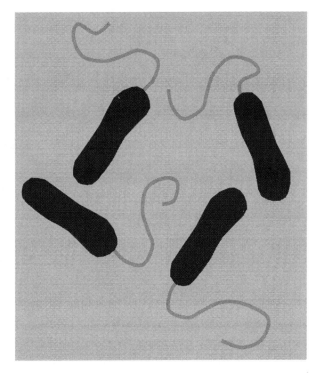

Figure 9-6   *Geobacteraceae bacterium.*

Here, oxygen is being bubbled through an electrolyte solution.

At the other electrode, we are bubbling oxygen through an electrolyte. The cathode electrode is losing electrons, therefore "loss equals oxidation."

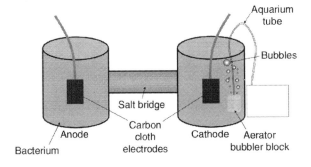

Figure 9-5   *Diagram of the microbial fuel cell.*

Loss Equals Oxidation
Gain Equals Reduction

Figure 9-7   *LEO the Lion goes GER—a great way to remember "Loss of Electrons Equals Oxidation" and "Gain Equals Reduction"*

Figure 9-8 *The reduction–oxidation cycle.*

So we can see how the reduction–oxidation cycle is complete, with one electrode reducing, while the other is oxidizing. The salt bridge provides a means for ions to flow between the two chambers while isolating the two solutions.

Mediatorless microbes transfer electrons directly to the electrode, without the need for an electron shuttle.

Other students' fuel cells

There are a plethora of good microbial fuel cell projects on the Internet. To all intents and purposes,

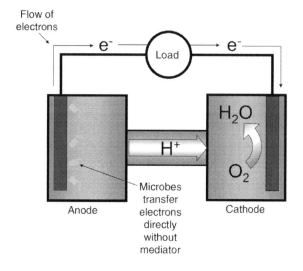

Figure 9-9 *Mediatorless transfer.*

microbial fuel cells make for very successful science fair projects, and being so new, there is exciting research that can be done by amateurs.

I *urge* you to take a look at Abbie Groff's award-winning microbial fuel cell research. When I began reading around this subject, Abbie Groff's work shaped a lot of my direction with this project:

www.geocities.com/abigail_groff/

Here are some of Abbie's press cuttings, resulting from her forays into the world of microbial fuel cells—discover something new and this could be you!

www.engr.psu.edu/ce/enve/mfc-Logan_files/PDFs/Student%20MFC%20papers/Abbie%20Groff4.pdf

Sikander Porter-Gill's project also makes for interesting reading:

www.engr.psu.edu/ce/enve/mfc-Logan_files/PDFs/Student%20MFC%20papers/MFC-Sikandar.pdf

Erik A. Zielke also has some interesting work:

www.engr.psu.edu/ce/enve/mfc-Logan_files/PDFs/Student%20MFC%20papers/Engr_499_final_zielke.pdf

Penn State University has a fantastic microbial fuel cells resource page (www.engr.psu.edu/ce/enve/mfc-Logan_files/mfc-Logan.html), and if you build an MFC like the one featured here, they want to hear from you! If you get a successful MFC project, send some details to blogan@psu.edu

Microbial fuel cells—with mediator

Mediatorless microbes transfer electrons directly to the electrode, without the need for an electron shuttle. The MFC we have created is a mediatorless microbial fuel cell because there isn't an additional chemical to "mediate" between microbes and electrodes (Figure 9-9).

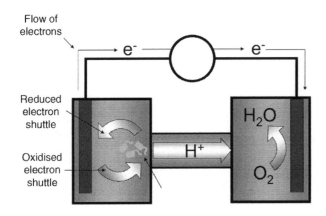

Flow of electrons

e⁻   e⁻

Reduced electron shuttle

Oxidised electron shuttle

H₂O

H⁺

O₂

Figure 9-10 *Mediator MFC.*

There is another type of microbial fuel cell—the mediator MFC, where an additional chemical acts to *shuttle* the electrons from the microbes to the MFC. This is illustrated in Figure 9-10.

Reading University's National Centre for Biotechnology Research has produced a very compact microbial fuel cell kit. The kit is based on the principles discussed in Peter Bennetto's paper (1990).

Look at Figure 9-10—this is what happens inside a fuel cell which requires a mediator. In the Reading University kit, yeast is used as the bacterium, and methylene blue is used as the mediator, shuttling electrons between the bacteria and the electrodes.

Reading University has published information about their microbial fuel cell in the online journal, *Bioscience-Explained*. You can download this information from www.bioscience-explained. org/EN1.1/pdf/FulcelEN.pdf

Further Investigation

- H.P. Bennetto (1990) Elecricity generation from micro-organisms: www.ncbe.reading. ac.uk/NCBE/MATERIALS/MICROBIOLOGY/ PDF/bennetto.pdf

- H. Liu and B.E. Logan (2004) Electricity generation using an air-cathode single chamber microbial fuel cell in the presence and absence of a proton exchange membrane: www.engr.psu. edu/ce/enve/publications/2004-Liu&Logan-ES&T.pdf

Chapter 10

High Temperature Fuel Cells

In this book so far, we have chosen experiments which can be performed with relative safety at home. The type of fuel cells that we have looked at can be classed as "low-temperature fuel cells"; however, there is another class of fuel cells that is suited towards different applications— "high-temperature fuel cells."

Take a look at Table 10-1. Here we can see the five main classes of fuel cells. Note that in this table, we group DMFCs with PEM fuel cells, as both operate on a membrane principle.

Unfortunately, because they operate at high temperature, it is not practical to construct or experiment with them, but we do at least owe them the duty to discuss them, so that your knowledge of fuel cell technology is complete.

High-temperature fuel cells can be used in stacks, like the stacks we saw in the PEM fuel cell. In Figure 10-1, we see a solid oxide fuel cell stack— the stack is made up of a sandwich of platters, a separator plate, followed by an anode plate, an electrolyte plate and a cathode plate. The sequence then repeats. Fuel is supplied through a central pipe, with additional air being supplied by the two pipes either side of the center.

Of course, because high-temperature fuel cells require additional support equipment, the inventory of components required to support a high-temperature fuel cell extends beyond the stack itself. We can see in Figure 10-2 a small 1-kW solid oxide fuel cell stack (the same one as we saw in Figure 10-1), in the lab with its associate support equipment.

Table 10-1
Comparison of fuel cell technologies.

	PEM fuel cell	Alkaline fuel cell	Phosphonic acid fuel cell	Molten carbonate fuel cell	Solid oxide fuel cell
Electrolyte	H^+ ions in a polymer membrane	OH^- ions in a KOH solution	H^+ ions in a H_3PO_4 solution	CO_3^- ions in a molten carbonate eutectic	O^- ions in a ceramic matrix with free oxide ions
Ability to reform fuels internally?				—	—
Oxidant: air	—	Purified	—	—	—
Oxidant: O_2	—	—			
Ideal operational temperature	65–85° C 150–180° F	90–260° C 190–500° F	190–210° C 370–410° F	650–700° C 1200–1300° F	700–1000° C 1350–1850° F
System efficiency	25%–35%	32%–40%	35%–45%	40%–50%	45%–55%
contaminated by:	CO, S, NH_3	CO, CO_2S, NH_3	CO, S	S	S

Figure 10-1 *Solid oxide fuel cell stack. Image courtesy Technology Management Inc.*

Figure 10-2 *Solid oxide fuel cell stack and support equipment in the lab. Image courtesy Technology Management Inc.*

Also, because of the high temperatures associated with high-temperature fuel cells (it's all in the name!), there is a certain amount of support equipment that means that for very small amounts of power, high-temperature fuel cells just aren't economical. To give you an example, high-temperature fuel cells don't get a lot smaller than the device shown in Figure 10-3, a 1-kW solid oxide fuel cell stack with support circuitry.

We are going to take a look at three high-temperature fuel cell types, and briefly cover some aspects of the technology.

Phosphoric acid fuel cells

In a phosphoric acid fuel cell (PAFC), the electrolyte is retained by a matrix of silicon carbide, which keeps the highly concentrated phosphoric acid in place.

Figure 10-3 *1-kW solid oxide fuel cell stack.*

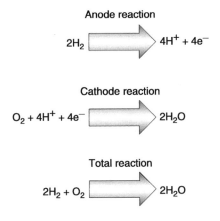

Anode reaction

$$2H_2 \longrightarrow 4H^+ + 4e^-$$

Cathode reaction

$$O_2 + 4H^+ + 4e^- \longrightarrow 2H_2O$$

Total reaction

$$2H_2 + O_2 \longrightarrow 2H_2O$$

Figure 10-4 *Phosphoric acid fuel cell chemistry.*

If you are still interested in phosphoric acid fuel cells, and want to find out more, you could check out some of these top Web links as a springboard to further information:

www.fctec.com/fctec_types_pafc.asp

www.nfcrc.uci.edu/EnergyTutorial/pafc.html

www.fossil.energy.gov/programs/powersystems/fuelcells/fuelscells_phosacid.html

americanhistory.si.edu/fuelcells/phos/pafcmain.htm

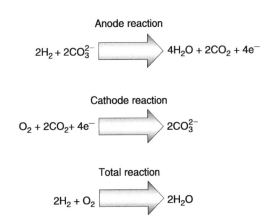

The reason for the high temperatures of phosphoric acid fuel cells is that at lower temperatures, phosphoric acid proves to be a poor conductor of ions. Furthermore, at lower temperatures, the fuel cell is more readily poisoned by carbon monoxide.

Also, phosphoric acid solidifies at 40° C or 104° F, which means that cold-starting of PAFCs can be rather tricky. To eliminate this operation complexity, PAFCs are usually used in applications, e.g., in buildings, where they can be installed and operate continuously without the need for cold-starting.

In addition to building integrated PAFCs, promise has also been shown in utility-scale applications where PAFCs can be used as mini power stations, producing supplemental power for the grid.

Additional heat produced from the PAFC operation must be removed. This can be done by incorporating coolant channels into the coolant stack—not unlike the coolant channels that run through the internal combustion engine in your car. Either liquid or gas can be passed through these channels to cool the PAFC.

Phosphoric acid fuel cells are one of the most mature fuel cell technologies. They are very well understood, and used in commercial operations at a number of sites.

Molten carbonate fuel cells

Molten carbonate fuel cells also operate at very high temperatures. Their electrolyte is a molten carbonate salt mixture, suspended on a ceramic matrix. In current MCFC fuel cells, two mixtures of carbonates are currently used: lithium carbonate and potassium carbonate, or lithium carbonate and sodium carbonate.

As a result of the high temperatures that MCFCs operate at, it is not necessary to use platinum as a catalyst; instead, other cheaper metals, such as nonprecious metals, can be used.

Anode reaction

$$2H_2 + 2CO_3^{2-} \longrightarrow 4H_2O + 2CO_2 + 4e^-$$

Cathode reaction

$$O_2 + 2CO_2 + 4e^- \longrightarrow 2CO_3^{2-}$$

Total reaction

$$2H_2 + O_2 \longrightarrow 2H_2O$$

Figure 10-5 *Molten carbonate fuel cell chemistry.*

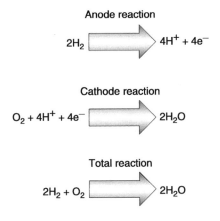

Molten carbonate fuel cells can reach very high efficiencies—even higher when waste heat produced by the fuel cell is captured. This heat can be used for other processes or heating, which means more useful energy is delivered to the application and less is wasted.

Other fuels can also be internally reformed, which presents flexibility in the range of fuels that can be used.

As a result of the high temperatures, MCFCs are resistant to poisoning by carbon monoxide and carbon dioxide. However, sulfur in fuels still presents a problem and must be scrubbed from the fuel before it is supplied to the MCFC.

As a result of the high temperatures required, these cells take some time to start up, and require some thermal shielding or insulation to be added to the cell to ensure that it operates efficiently.

One of the problems that needs to be surmounted at the moment is the durability of MCFCs: some of the materials that they use in the electrolyte are highly corrosive, and as a result of this, and the fact that these cells are operating at a *very* high temperature, premature component failure can result.

www.

If you want to learn more about molten carbonate fuel cells, the following are some super links to sites with high-quality content that you might like to check out:

www.fctec.com/fctec_types_mcfc.asp

www.fuelcellmarkets.com/fuel_cell_markets/mol ten _carbonate_fuel_cells_mcfc/

www.fossil.energy.gov/programs/powersystems/ fuelcells/fuelcells_moltencarb.html

en.wikipedia.org/wiki/Molten-carbonate_fuel_cell

dodfuelcell.cecer.army.mil/molten.html

americanhistory.si.edu/fuelcells/mc/mcfcmain.htm

www.nfcrc.uci.edu/EnergyTutorial/mcfc.html

Solid oxide fuel cells

As a result of the very high temperatures that they operate at, solid oxide fuel cells are primarily suited towards stationary applications. The gases that are produced by SOFCs are so hot that, on their own, they can be used to power a gas turbine. Further low-grade heat can then be recovered after the turbine for heating purposes.

SOFCs must run at *very* high temperatures, as the materials they are made from only become electronically and ionically active at *very* high temperatures.

Like the MCFCs, because of the high temperatures required, these cells take some time to start up, and require some thermal shielding or insulation to be added to the cell to ensure that it operates efficiently.

As a result of the very high temperatures that SOFCs operate at, poisoning of the fuel cell by carbon monoxide is not an issue. However, sulfur must be scrubbed from the fuel supply as SOFCs are still susceptible to sulfur poisoning.

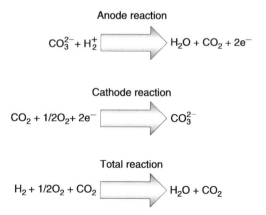

Anode reaction

$$CO_3^{2-} + H_2^+ \longrightarrow H_2O + CO_2 + 2e^-$$

Cathode reaction

$$CO_2 + 1/2 O_2 + 2e^- \longrightarrow CO_3^{2-}$$

Total reaction

$$H_2 + 1/2 O_2 + CO_2 \longrightarrow H_2O + CO_2$$

Figure 10-6 *Solid oxide fuel cell chemistry.*

As a result of the solid oxide fuel cell's tolerance to poisoning, and its ability to reform fuels within the fuel cell, it is suited to a wide range of different hydrogen-containing fuel sources.

Shared with the molten carbonate fuel cell, one of the advantages of the solid oxide fuel cell is that it can *reform* fuels in situ, which means that we can supply it with a supply of natural gas direct from the gas pipeline, and it will reform the gas itself, with no need for an intermediate stage. This provides additional flexibility for some applications.

There is a wealth of material about solid oxide fuel cells on the Web. Here are some links to pages with high-quality information:

www.svec.uh.edu/SOFC.html

www.fossil.energy.gov/programs/powersystems/fuelcells/fuelcells_solidoxide.html

www.iit.edu/~smart/garrear/fuelcells.htm

www.azom.com/details.asp?ArticleID=919

Chapter 11

Scratch-Built Fuel Cells

If you have got this far in the book, the chances are you now have the armory of knowledge required to start thinking about building your own fuel cells from scratch. In this chapter, we are going to look at some of the skills required to build membrane fuel cells.

If you haven't already had the chance, take a peek at the chapter on direct methanol fuel cells (Chapter 8), and the project "Build Your Own Band-Aid® Fuel Cell." It was in devizing this project that I began to realize how simple fuel cells really are—simply a sandwich of components in the right order, made from the right materials.

Although the construction is fairly simple, the skill of manufacturing fuel cells is to ensure that all components are optimized—this is the challenge of manufacturers, who every day are working on ways to make fuel cells smaller and cheaper.

In this chapter, we are going to briefly explore some of the tools, techniques, and materials of scratch-built fuel cell fabrication. The ease of building your own fuel cells from scratch is that many of the components are available as premade off-the-shelf items, if you don't want to *totally* start from scratch, but as you gain skills, experience, and confidence, you can begin to build more components and vary the balance between off-the-shelf and homemade—to the point where you can then start from raw materials and turn out commercial-grade fuel cells.

We are going to start at the ends of our fuel cell and work our way in.

It is possible to make fuel cells in a variety of shapes and sizes, however, there are some components available which follow a standard sizing, and have holes and membrane areas in

common places. Until you get confidence with building fuel cells, it's probably a good idea to start from one of these template layouts, using stock components to begin with, and then fabricating new components as you gain in confidence. This has the benefit of you being able to take dimensions from the old components as patterns, modifying parameters one by one, and seeing what effect it has on the fuel cell's performance.

Scratch-built fuel cell components

End-plates

The job of the end-plates in a fuel cell is to be rigid enough to firmly support the internal components of the fuel cell, to provide a fixing for the hydrogen and oxygen supply connections, and to provide a surface for the fixings that secure the fuel cell sandwich together to screw down on.

A wide variety of materials are suitable for end-plate construction. You might want to consider using an electrically insulating material for the end-plate, to ensure that it doesn't provide a path for stray electrons to short on. Fiberglass sheet is an ideal material to produce end-plates from, as it is strong, robust, and can be easily worked. Some plastics are suitable for end-plate construction, however, you must ensure that they are sufficiently rigid, as a material with too much bend will not provide the mechanical support the MEA needs through the center of the stack.

Figure 11-1 *Fiberglass sheet suitable for end-plate fabrication.*

You want to ensure that the material does not bend significantly, as you want pressure to be spread from the screws or clamps at the outside of the cell, to the center of the cell, where the end-plates maintain the electrodes' contact with the MEA.

Hint

If producing your end-plates sounds like too much trouble, Fuel Cell Store sell a set of end-plates with mounting components for a 10 cm × 10 cm fuel cell stack–Part No: 590300.

To produce an end-plate, start with a fiberglass sheet, such as that shown in Figure 11-1. The Fuel Cell Store sells a garolite (fiberglass and epoxy) sheet that is suitable for end-plate construction. Their stock number is 72043001.

The end-plate needs to have holes for nuts and bolts to be inserted through, which are then tightened to provide pressure on the stack. Another alternative is to use a threaded rod with nuts on either end.

The end-plate also needs to have a suitable fixing for hydrogen and oxygen supplies and return feeds. A simpler construction of fuel cell will use hydrogen and air, in which case pipes for oxygen do not need to be provided. If you think back to the fuel cells you have been using in the chapter on PEM fuel cells, they all have their integral fixings molded as part of the plastic end-plate. However, if you are working from a fiberglass sheet, you will need to find a way to affix a connection to the plate.

You will see that the end-plate in Figure 11-2 has four holes at the corners for fixings, and two connections for hydrogen and return gas at the top in the center of the plate.

The positioning of your holes and gas feeds can differ slightly—this is just an example. As end-plates are one of the easier parts of the fuel cell stack to manufacture, it is probably best to design them around any available flow fields, rather than the other way around.

If you haven't got the facilities, or you haven't got the patience to sit and make your own end-plates, the Fuel Cell Store sells some ready-manufactured end-plates, with holes in the right

Figure 11-2 *An example end-plate.*

Figure 11-3 *Gas connection.*

places for their ready-made flow fields. The end-plate kit comes with barbs for gas already fitted, and a set of eight insulated fixings to hold the fuel cell together. The stock number is 590300.

There are a number of different fixings for gas connections. It is important that you can make a good leak-proof joint between the "barb"—the bit that sticks out to interface with the pipe, and the end-plate. Some barbs will simply screw into the end-plate, while others may require some epoxy or other form of adhesive.

Graphite flow fields

The next component inbetween your end-plates is the flow fields. It is the flow fields that take the gases from the common feeds machined throughout your fuel cell stack, and distribute them evenly over the MEA. In addition to the mechanical function of helping distribute the gas, and providing physical support to the MEA, they also have the electrical function of providing a contact to the gas diffusion membrane to allow the power to flow!

You need to make a decision as to whether your fuel cell is to run on hydrogen and air, or hydrogen and oxygen.

The hydrogen distribution design remains similar for both types of fuel cell—you take the hydrogen from a common point which feeds from the barbed tube connector at the end of the fuel cell, and using a flow field channeled in the graphite, you provide a passage for the hydrogen to spread over the surface of the MEA. You can see an example hydrogen flow field plate in Figure 11-4.

Notice how there are four holes on the outside of the plate at each corner: these are to provide the manner of fixing the fuel cell together, with bolts passing through all the plates and the entire assembly. Meanwhile, you will notice two small holes at the top of the plate—these are to connect the serpentine pattern, which we can see below, to two common channels which provide connection to the hydrogen feed.

If your fuel cell is to run on hydrogen and oxygen, you will want to take a similar pattern to Figure 11-4, but with the holes positioned in a different place for an oxygen feed. This will then take a supply of oxygen, attached to a barb fixed in the opposite end-plate in a different position, and distribute the oxygen over the opposite side of the MEA to the hydrogen.

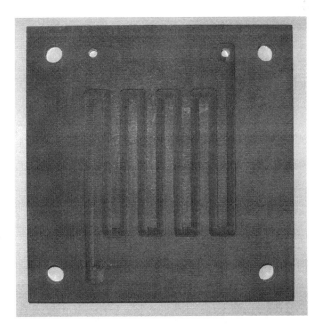

Figure 11-4 *Hydrogen flow field plate.*

Figure 11-5 *Flow field plate suitable for a hydrogen-air fuel cell.*

If your fuel cell is to run on hydrogen and air, you will use straight plates on the "oxygen side" to allow air from outside the fuel cell to flow along the straight channels to make contact with the MEA. There are no tubes, the air enters the fuel cell at the bottom and leaves at the top. Natural convection currents created by the generation of any excess heat can help to increase the circulation of air. A straight plate is shown in Figure 11-5. Additionally, you could experiment with forced ventilation, attaching a fan to one side of the stack to increase the throughput of air.

Hint

If you want some ideas for different patterns for your flow fields, take a peek back at Figures 7-10 for straight, Figure 7-11 for interdigitated, Figure 7-12 for serpentine, and Figure 7-13 for spiral—for inspiration.

You can buy ready-machined flow field plates which take the hassle out of making your own. They are available from Fuel Cell Store—this Web link will show you all their ready-made plates: www.fuelcellstore. com/cgi-bin/tornado/view=NavPage/cat=58.

If you are building a single-cell fuel cell, you will require one anode plate, with channels on one side, and one cathode plate, with channels on one side. However, if you are making a fuel cell stack, you will need to employ "bipolar plates." A bipolar plate has the channels cut out on one side for hydrogen and the channels cut out on the other side for oxygen (or air). It acts as an electrical conductor, allowing power to be transmitted down the length of the fuel cell stack, while also providing a method to keep the hydrogen and oxygen gases separate. The bipolar plates fit in between the end anode plate and the end cathode plate.

If you are looking at producing your own flow field plates, you will need to start with a plain carbon block, cut to the dimensions of your fuel cell. In terms of thickness, you need to allow enough material to be moved when you are milling the channels for the gas to flow through. If you look at Figure 11-6, you will see a plain carbon plate suitable for the purpose.

You can buy blanks ready for your own machining and treatment, or you can cut your own from a block of graphite. Before cutting slices, square (or rectangle) from your raw block of graphite, ensure that you check the sides for parallelism.

Graphite is a lovely material to work with—it's really nice and soft to mill into, so you don't need especially flash equipment to cut the flow channels into it. Consider when working with your graphite that a little bit of water will aid the cutting process and keep the workpiece cool. Graphite is a good lubricant, so you won't need cutting fluid—just something to keep the dust down and remove the heat.

When drilling or milling the graphite plate, set your drill or mill to the slowest speed possible.

Figure 11-6 *Plain graphite plate suitable for flow field construction.*

You do not need to work at high speed when machining graphite.

When considering tools, you have a number of options. Probably the simplest is to use one of the handheld multifunction tools, such as the Dremel Multi, which offer a wide range of heads and attachments for different purposes. This will enable you to manually route out a groove in the carbon—you can then drill your holes using a small bench-top pillar drill.

If you've got access to a more sophisticated workshop, you could consider using a milling machine, to enable you to make accurate straight lines for your serpentine pattern. For those of you who are now lost, with a milling machine, you can clamp the workpiece on a flat bed, and you have a number of controls—if you think of the old Etch A Sketch machines, you have two knobs: an X and a Y, and they allow you to draw lines with the cutting tool. The difference is that you can also start and stop the line by raising the milling bit, Z.

Now, this is all well and good for single flow fields; however, if you are building a fuel cell stack, the chances are you will quickly find this becomes a tiresome exercise—there's also a distinct possibility that if making the flow-fields manually, there will be some operator error then, and they might not all line up correctly.

The *pièce de resistance* of manufacturing flow fields is to invest in a CNC cutting machine. CNC stands for computer numerical control—it's the lazy man's solution, as it allows you to create a design in a computer-aided design package, and *send* it to the machine, which will then mill the material as dictated by the pattern.

Increasingly, you will find CNC machines in well-equipped schools and college design and technology departments, so if you're friendly with a teacher or technician, you may be able to manufacture some flow fields.

Alternatively, there are some fairly simple CNC kits, such as from Parallax, which is a simple CNC machine for light hobby work. It is eminently suited for this kind of work, which is very lightweight. The machine is much more affordable, by a number of orders of magnitude, than commercial CNC machines, and will find other applications in an Evil Genius's workshop!

You need to ensure a good interface between the graphite flow field and the MEA, so you want to ensure that the surface of your flow field is as smooth as possible. Wet and dry paper is a good way to remove imperfections—start with coarser grits working down to a 600 grit. One thing is essential—work with a flat-bottomed sanding block, as it is absolutely critical that parallelism is maintained on either side of the flow field plate.

Membrane electrode assemblies

In a membrane-based fuel cell, it is the assembly of membrane as gas diffusion electrodes that you will understand does all of the work. It is at the junction between membrane, platinum, and electrode

Figure 11-7 *An off-the-shelf MEA.*

Figure 11-8 *Always remember to handle MEAs with cotton gloves.*

that the chemical reactions vital to the fuel cell's operation take place. We saw in the chapter on PEM fuel cells how this happens.

It is possible to buy the MEA as a ready-fabricated component, however, by making your own, you gain a greater degree of control over the materials used in the performance of your MEA, you can make custom sizes, and are able to produce MEAs at cheaper cost. Also, by making your own, it is possible to have much greater control over the coating process—this additional flexibility gives wider scope for experimentation and new learning.

The chemical coating of an MEA is very sensitive to pollutants and very fragile. For this reason, when handling MEAs—shop bought, or making your own—always ensure that you wear fine, clean cotton gloves as they ensure that you will not damage the MEA. Handle them carefully, as they are very delicate, and try and hold them around the edges if possible, rather than making contact with the black central portion.

Project 32: Making Your Own MEAs

For the hardcore fuel cell experimenter, you might want to have a go at making your own MEAs. In order to do this, you are going to need some pretty sophisticated materials and some relatively cheap tools.

To recap for those that have forgotten our little journey down PEM avenue, MEA stands for Membrane Electrode Assembly.

An MEA consists of a proton exchange membrane, Nafion™ from Du Pont is commonly used. This is sandwiched between either a cloth or paper impregnated with carbon. This forms our GDL or gas diffusion layer. To recap, the GDL allows our gas and water to move to and from the platinum catalyst.

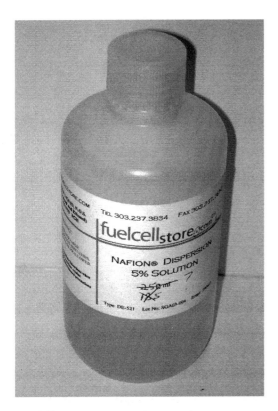

Figure 11-9 *A bottle of Nafion™ solution.*

We will be using Nafion™ solution in the construction of our own MEAs—this can be purchased in a variety of percentage solutions.

You will also need some Nafion™ membrane in sheet form. Remember, when it arrives in the

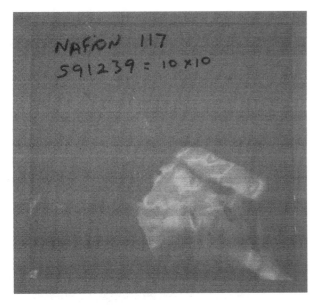

Figure 11-10 *A sample of Nafion™ membrane.*

post, resist the urge to get it out and start playing with it, as it is easily damaged.

You can obtain both Nafion sheet, and Nafion membrane from Fuel Cell Store—this website link takes you to all their Nafion™ products: www.fuelcellstore.com/cgi-bin/tornado/view=NavPage/cat=97

Warning

If you can do this at a school or college chemistry lab, then use a fume cupboard—this will ensure that you are not harmed by the vapors from the boiling hydrogen peroxide or sulfuric acid. However, if you do not have a fume cupboard, ensure that you do this activity outside in a well-ventilated area to ensure your safety. Make sure that you wear protective clothing while carrying out this activity—some thick acid-proof gloves or gauntlets will ensure that your hands and arms don't come into contact with boiling acid. Ensure that you are wearing safety specs, use a protective apron, and ensure you are wearing fairly thick clothing to cover up your arms and legs.

In the event of a splash, turn off the heat, and go indoors immediately, take the affected piece of clothing off, and use water to make sure that no chemical is in contact with your skin for too long.

You will need to *prepare* your Nafion membrane for use in your fuel cell. You will need a double-boiler, sometimes called a Bain Marie, which you will not be using again for food, as we're going to be using some nasty chemicals in it.

You can make a Bain Marie simply from laboratory glassware—it just consists of a larger pot which can be heated to the point of water boiling, with a smaller pot inside which is heated indirectly by the hot water. The intermediate step of heating the water gives you finer control over the temperature, and a more even distribution of heat, which is useful in this application.

It is also handy if you have a setup where you can heat two pots at the same time, as this will allow you to begin to bring another bath up to temperature, while the one the MEA is currently in is coming to the end of its cycle.

Warning: a metal Bain Marie is not suitable for this experiment—you need to ensure that you are using glass or something inert which will not react with the hydrogen peroxide or sulfuric acid.

You will also need a thermometer capable of measuring relatively high temperatures—don't be tempted to use a standard glass thermometer—dip the end in the hot, boiling water, and you will certainly have damaged your thermometer—it might even crack, leaving the mercury or alcohol to leak out. One place you could start your search for high-temperature thermometers is online, looking at supply stores for home sweet or candy makers.

You are going to need to cut a uniformly sized piece of Nafion™ for each MEA in your fuel cell stack.

Cut your carbon cloth or paper, so that it is slightly smaller than the Nafion™ membrane, giving it a border which can be used to mount the membrane as part of a larger sheet of gasket.

To prepare the membrane, you will need 3% hydrogen peroxide solution, sulfuric acid, and plenty of distilled water.

We will be filling the larger pan of the Bain Marie with water, which we will bring to the boil. The smaller pan which sits inside the pan will start off with distilled water. You need to immerse the MEA in this water, at a temperature of 176° F or 80° C to hydrate the membrane for a period of around an hour. Use a glass rod to ensure that the MEA is

fully immersed in the water, as its tendency will be to float up.

Towards the end of the hour, you need to bring another pot—this time with 3% peroxide solution—up to temperature (176° F or 80° C again). Transfer the membrane from one pot to the next, and maintain the temperature for an hour.

The membrane may be tricky to transfer—being transparent it's hard to see, and you'll probably find it slippery to handle. There are three golden rules:

1. Don't drop the membrane.
2. Don't puncture or scratch the membrane.
3. Don't use a metal object to retrieve the membrane.

Take the pot you used for the initial distilled water bath, and transfer a quantity of sulfuric acid into this pot. Bring it up to the temperature of 176° F or 80° C.

Take the membrane out of the hydrogen peroxide, and allow as much of the fluid to drip from the membrane as you can, and then immerse the membrane in the sulfuric acid. Use the glass rod again to ensure that the membrane sits below the surface of the hot liquid. Once this has been heated for an hour, we will be *washing* the membrane in three distilled water baths at 176° F or 80° C. Once this process is complete, allow the membrane to dry—ensure that it doesn't come into contact with anything metallic, and ensure that the air can get to it.

Warning

Remember when you are finished with your sulfuric acid to dispose of it in a sensible, responsible manner. Do not store it in any kind of generic container that could be mistaken for a food item or cleaning product—make sure you store it responsibly in a clearly labeled container.

Preparing the GDL

Remember the gas diffusion layer? Well, you're going to have a choice of either using carbon paper, as shown in Figure 11-11, or carbon cloth, as shown in Figure 11-12.

The gas diffusion layer is a carbon impregnated cloth or paper, with a high concentration of platinum which acts as a catalyst. For the home fuel cell experimenter, there are a number of ways that we can prepare the GDL.

For a range of GDL materials, check out Fuel Cell Store's page at: www.fuelcellstore.com/cgi-bin/tornado/view=NavPage/cat=56

We're going to discuss a few different options for transferring the catalyst to the carbon cloth or paper.

You're also going to need some platinum catalyst to deposit on the interface between your GDL and the Nafion™ membrane. In this respect, you have an array of options open to you. It is quite handy

Figure 11-12   *A sample of e-Tek gas diffusion layer.*

and easy to buy ready-prepared mixtures of carbon and platinum powder for your catalyst. This can be applied to your GDL. You can get jars of anode and cathode paste with varying compositions of catalysts, which can be applied to your GDL.

Remember, for repeatability, always make fine measurements of the quantities of chemicals that you are using, so that you can manufacture similar fuel cells in the future. Compare different loadings of platinum catalyst to see how it affects the performance of your fuel cell.

Figure 11-11   *A sample of carbon paper.*

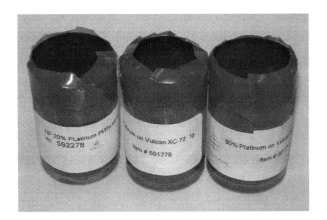

Figure 11-13   *Jars of anode and cathode pastes.*

Figure 11-14 *Fine scales for carefully measuring chemical quantities.*

High-voltage deposition

One way in that it is possible to transfer platinum to the GDL is through high-voltage vaporization of fine platinum leaf. By passing a very high voltage through a fine wire or leaf of platinum, it is possible to vaporize it, and small pieces of platinum will be deposited on any adjacent surfaces that are close by.

There are a couple of books that discuss making devices to vaporize fine wire in detail:

Two books that I can highly recommend:

*Gordon Mc Comb's Gadgeteers Goldmine—*Gordon Mc Comb, McGraw-Hill, New York

*Electronic Gadgets for the Evil Genius—*Robert Ianinni, McGraw-Hill, New York

Platinotype—photochemical deposition

Platinum is firmly embedded in the history of photography, and as such, platinum printing, while being expensive, does furnish the potential fuel cell builder with a method of accurately depositing measured amounts of platinum compound into your MEA.

The platinum printing process was developed by W. Willis in 1873. For fuel cells, it is perfect as it deposits a layer of platinum black onto the MEA—this is especially useful as platinum black is one of the best forms of platinum for our application.

You won't find platinum printing kits in your standard photo shop, as the process is now largely used for art prints, or prints with a very high resale value. However, there are a couple of online vendors that sell the chemicals for this process.

Here is a great source of platinum printing kits, and all the chemicals you need to platinum print are also available individually:
www.photographersformulary.com

Follow the instructions in the kit (which may differ depending on the supplier). You will need to coat your pieces of GDL as if they were photographic paper, but rather than enlarging an image onto them, I suggest:

- Leaving them out in the sun for UV exposure

- If you have a UV light box (say for etching printed circuit boards), you can use this to expose the piece of GDL.

Remember, photographic paper takes a negative image, and reverses it, so for areas of high-density black, we need to expose the GDL to bright white light. Once you've made the exposure, don't forget that you'll need to *process* your GDL using the chemicals in the kit, in order to *fix* the platinum black.

Throughout the processing, keep a keen eye on what side you applied the platinum to, as the fuel cell will only work if the platinum is in contact with the Nafion™ membrane.

Electroplating

Another option available to you is to electroplate platinum onto your GDL. To do this, you connect the GDL to the negative terminal of a power supply, a piece of platinum scrap to the positive, and immerse the two objects in a solution of platinum salts.

www.caswellplating.com sells a wide range of home electroplating kits and accessories; however, they do not supply kits specifically for platinum plating. Their website has a useful forum, with lots of information about electroplating with different metals, and this would be a good place to start if you want to investigate platinum electroplating.

If you have bought the above-mentioned platino-type process kit, you should have some potassium tetrachloroplatinate included in the kit—this is a platinum salt, and all you need now is some platinum for the anode, which could be a piece of platinum leaf.

Platinum leaf

You can transfer platinum leaf very easily to carbon cloth or paper, by simply holding the leaf against the GDL and rubbing hard on the back with a pointed but not sharp object (e.g., the wrong end of a ballpoint pen).

Putting it all together ...

Once you've impregnated your GDL with platinum, and prepared your Nafion™ membrane, the next step is to assemble your MEA.

To do this, you are going to need a couple of stiff sheets of metal and some clamps to apply pressure between the two sheets. You might find that you can get a ready-made assembly from a paper-making kit—the key is, you need to ensure that the materials are happy being heated in the oven.

The Fuel Cell Store sells a pair of aluminum plates suitable for hot-pressing your MEA—the stock number is 590163.

You will now need to coat your GDL with some Nafion™ solution. Remember Figure 11-9? Ensure that you do not get any metal in contact with the Nafion™—you can get cheap plastic brushes for coating pastries with egg or kiddies' art-brushes. Just coat one side of the GDL (the side with the platinum on)—you don't need to saturate the carbon cloth or paper, but just ensure that the surface is coated.

You're also going to need an oven thermometer, of the type you use to check that your turkey is at the right temperature.

Remember that we said earlier what a fantastic lubricant graphite was—well, we're going to need some graphite powder now (see Figure 11-16), to ensure that the Nafion™ doesn't stick to the plates we are using to clamp the assembly together. This is just like greasing a pan when we are cooking. Rub the graphite into the surfaces of the plates,

Figure 11-15 *Graphite powder.*

Figure 11-16 *Home-built fuel cell (note fixings and tube barbs).*

which will be in contact with the clamped MEA assembly.

Graphite powder is available in small bottles from Fuel Cell Store—the stock number is 590263.

We're now going to assemble the MEA—this is a bit like putting a burger together. First, place the first piece of GDL platinum side up in the center of one of the plates. Lay the Nafion™ membrane over this, ensuring that it overlaps on all sides, and then lay the final piece of GDL on top, platinum side down. Carefully place the final plate on top, graphite side in contact with the piece of GDL, and firmly clamp this assembly together.

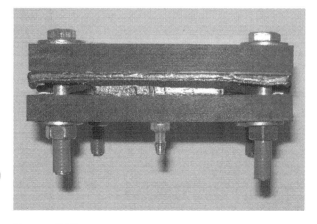

Figure 11-17 *Home-built fuel cell, side view.*

Figure 11-18 *Home-built fuel cell (note electrodes and sandwich construction).*

Now bake …

Using the oven thermometer to check the temperature is accurate (never rely on the temperature indications on your oven dial—they're not accurate enough), bake at 194° F or 90° C for an hour, and then slowly over the course of the next half an hour, turn the knob on your oven up to 266° F or 130° C. After the sandwich has been in for an hour and a half, whip the lot out, tighten the clamps even more (using tea towels and oven gloves to avoid serious burns). Now, put the sandwich back in your oven, and whack the temperature control up to its maximum. Watch the oven thermometer carefully—we want to get the oven up to 266° F or 130° C, but no more! Maintain this temperature for a couple of minutes. Then switch your oven off, remove the clamped assembly, and leave it to cool on a trivet until it reaches ambient temperature.

Gaskets

You will need to mount your MEA in a gasket, which will ensure a good interface between the MEA and adjacent flow fields, and will also prevent it from diffusing from one side to the other. Look for a thin silicone rubber sheet—you will need to cut two pieces, one for either side of the MEA, and use a silicone glue to run a thin bead of adhesive around the overlapping piece of

Nafion™ membrane. Remember, your gasket will need a hole in the middle which is the same size as your GDL, holes in the corners to allow the nuts and bolts or threaded rods to pass through, and holes where the hydrogen (and oxygen if present) gas supply lines are.

Fuel Cell Store sells a range of different materials that can be used to manufacture gaskets for your MEA. Look on their website at www.fuelcellstore.com/cgi-bin/tornado/view=NavPage/cat=62, and you will see that they sell Teflon, rubber, silicone and Mylar—all suitable materials for different fuel cell gasketing applications.

Electrodes

You're going to need a metal electrode at either end of your fuel cell assembly to tap off the power. There are a couple of ways of doing this. You can use metal end-plates, however, this requires that you pay great attention to insulating the fixings used to hold your fuel cell together, or you can simply insert a sheet of metal between the end-plate and the flow field at each end.

Fixing it all together

You are going to need some fixings to hold your fuel cell together—you can either use nuts and bolts, or threaded rods. However, the key is to remember to insulate the portion of the thread inside the fuel cell, to prevent it shorting electrodes and flow fields. This can be simply accomplished with some heat-shrink tubing.

Further reading

- Hurley, P., *Build Your Own Fuel Cells*, Wheelock Mountain Publications Wheelock, VT.

This is an invaluable reference for those wishing to take this even further. It expands upon methods of platinum deposition and constructing MEAs in much more detail than is within the scope of this book, and is a mine of useful information for further research.

Chapter 12

Hydrogen Safety

Dangerous hydrogen? Safety concerns or flimsy myth?

As with any technology, there is generally a fair share of urban myths and legends that circulate, which aren't true, don't give us useful information, and end up cultivating a distrust or fear of a certain technology.

It is the same with hydrogen and fuel cells. In this chapter, we are going to debunk and demystify some of the half-truths surrounding hydrogen, and restore some confidence in the technology. This is going to be followed by a coverage of some of the health and safety aspects you should be aware of when experimenting with hydrogen.

Hydrogen is safe, but with any form of chemical or experimentation, it is important to treat all procedures with due care and respect.

Hydrogen has got a bit of a bad press from many different sources. For many years, graceful, lighter-than-air airships, filled with hydrogen gas, graced the skies. Following the Hindenburg disaster, the airship dropped into relative obscurity. The hydrogen gas that it was filled with is often blamed as the culprit, but the hydrogen burned upward. In fact, this is one of the inherent safety features of hydrogen. Because hydrogen burns "upwards" it would not have affected the canvas below; however, it has been proved that it was in fact the silver coating of the ship that caused the most damage—this coating was covered with powdered aluminum. Testament to this was the fact that out of 37 casualties, 35 died from jumping to the ground from

a height. Diesel burns were also responsible for a lot of injury.

The fact that hydrogen is so light means that when it is lit, it burns "up and away" from you. Contrast this to say, petrol, where fumes are heavy, and creep along the ground—the number of car mechanics who have received injury from fires igniting petrol vapors that have slowly accumulated in pits is testament to this. Whereas with hydrogen, so long as there is sufficient ventilation, and room for it to escape, it will rise up and dissipate. However, when designing buildings to house hydrogen equipment they must not have a "sealed cap," as rising hydrogen can accumulate here.

It's easy to read into the name of "the hydrogen bomb" the notion that hydrogen is dangerous. We have all seen the devastating mushroom clouds that nuclear weapons leave in their wake. Since then, the hydrogen bomb has done nothing for the good name of hydrogen, linking a potential savior of our energy future with dangerous nuclear technology. The hydrogen bomb contained tritium, radically different from the hydrogen that would be used in the hydrogen economy. Tritium is an isotope of hydrogen, and the sophistication and technology required to make an atomic bomb just isn't accessible. The mechanism that a hydrogen bomb uses to create its massive devastation involves the fusion of hydrogen atoms, something which is quite different from the chemical processes that take place when burning hydrogen, and so the two should not be confused.

Furthermore, a number of accidents in space have become linked in the public mind with hydrogen—this just isn't fair!

The Apollo 13 mission in space seems to have mysteriously become associated with hydrogen in

Figure 12-1   *The Hindenburg disaster.*

many people's minds—this is another urban myth. In 1970, on April 11th, NASA launched the 13th Apollo mission. During the flight, some fans inside one of the ship's oxygen tanks were turned on, and there was a short circuit, setting the fan on fire. As a result, the oxygen tanks were weakened

Figure 12-2   *The hydrogen bomb.*

and explosive. I don't know how much you know about astronauts, but they don't get on so well without a supply of oxygen. Luckily, mission control were able to work out a solution for the astronauts in space, and in true prototype experiment fashion, duct tape was employed to stick things together. (See Figure 12-3.) Contrary to popular belief, hydrogen and fuel cells were not in any way related to this disaster.

While the Apollo 13 mission can be shown not to cast a slur on hydrogen or fuel cell technology, the mission does provide a fantastic endorsement for duct tape (a marvelous product!) which saved the day at the last minute by assisting the astronauts in making a jerry-rig fix, which safely returned them home. Duct tape is long overdue a presidential Medal of Freedom for outstanding service to the world!

The Challenger disaster in 1986 also linked hydrogen with safety, and has led many to believe that hydrogen technologies cannot be safe. However, NASA have concluded that the disaster was not

Figure 12-3 *Jerry-rigged solution which saved the astronauts of Apollo 13.*

caused in any way by hydrogen, instead it was caused by faulty "O-rings."

Fortunately, all of these instances do not relate to how hydrogen would be used in a hydrogen economy, and any speculation arising from these events can be discounted as hysterical overreaction.

Hydrogen safety in vehicles

Any fuel which embodies a lot of energy in a small space has the potential to cause danger. If we fill up a car with petrol, we take a calculated risk, but everything is done to reduce that risk. Filling a car with hydrogen would involve no more risk.

One of the things about hydrogen is that it is lighter than air. This means that if you spill some, rather than landing on the ground, it disperses and floats to the sky. This is quite an advantage, when you come to consider how much petrol and diesel has polluted our watercourses. Furthermore, you won't stain your clothes with hydrogen as it just disperses.

There is a lot of work going into developing strong, robust hydrogen storage tanks that, in the case of vehicles, can withstand even the most catastrophic crashes. Sure, hydrogen has the potential to explode, but drivers of petrol vehicles have happily exposed themselves to the same risks for many decades. Following extensive crash testing, BMW concluded that in a 55-mph impact, hydrogen vehicles are no more dangerous than carbon-based vehicles.

One of the considerations that manufacturers take into account is that there can be no accumulation of hydrogen in the vehicle should a leak develop.

But what if hydrogen leaks from your car? Research has shown that if hydrogen leaks from a vehicle tank at a rate of 3000 cubic feet per minute and is lit, the temperature inside the vehicle does not climb any more than 1 or 2 degrees—vehicles get hotter than this sitting in the sun!

This is because hydrogen produces little radiant heat. When a carbon-based fuel burns, particles of hot carbon (soot) transfer heat to the surroundings—with hydrogen there is no soot and so less radiant heat transferred.

Hydrogen is already used extensively in industry, where safe methods for the transportation and storage of hydrogen have been developed. The recent Buncefield Oil Depot disaster in the UK, and countless other accidents around the world involving oil, have highlighted the dangerous nature and shortcomings of storing hydrocarbon fuels—storing hydrogen would be no more dangerous.

One of the problems with hydrogen is that it burns with a colorless flame—this means that we can't always clearly see a hydrogen fire. If you suspect a hydrogen fire, do not try to approach it, and do not try to identify it by "feeling" heat. The correct method is to take a long-handled broom, and waft the bristles of the broom around where you suspect there is a hydrogen flame. If hydrogen is burning, the bristles on the end

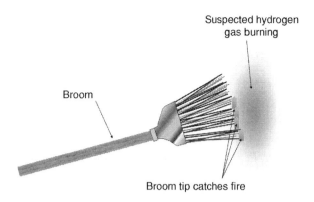

Broom

Suspected hydrogen gas burning

Broom tip catches fire

Figure 12-4   *Identifying a hydrogen fire with a broom.*

of the brush will catch light (see Figure 12-4). If you identify a fire, quickly pull the broom away, and call the fire brigade. If you can safely isolate the supply of gas, without getting anywhere near the flame, for example, by a remote valve, or gas shutoff safety system, do so.

Running your experiments safely

In order to avoid accidents and run a safe experimentation area, if you have any stored flammable gas, in bottles, and/or hydrides, make sure that they are stored correctly and labeled with a suitable label, such as that shown in Figure 12-5.

Make sure that whenever you are carrying out experiments with hydrogen, methanol, or other gases or fuels, that you are working in an area that is well ventilated to allow any gases that are produced to escape, preventing build-up, and explosion hazards.

Make sure you keep all bottles of chemicals clearly labeled and never assume that the label is right. Always treat chemicals as an "unknown quantity"—a clear liquid could be tap water, but then it could also be concentrated acid!

Warning

If you are working with the small quantities of hydrogen produced by a little PEM electrolyzer, this isn't going to happen, but if you start using hydrogen from a cylinder, or from a laboratory tap, or working with larger quantities of hydrogen without proper ventilation, beware!

In the event that a person is asphyxiated by inhalation of hydrogen gas:

The person should be moved to an area of ordinary atmosphere, ensuring that you do not endanger yourself in the process. Open windows, provide as much ventilation and access to air as possible, while making sure not to switch on any electrical appliances or cause any sources of ignition.

FLAMMABLE GAS

2

Figure 12-5   *Clear labeling is essential.*

Figure 12-6   *No smoking.*

Figure 12-7 *Lab coat.*

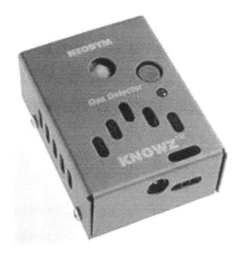

Figure 12-9 *A hydrogen safety sensor. Image courtesy Fuelcellstore.com.*

Finally, something that should seem incredibly obvious: there should be no smoking. Reinforce this message by photocopying Figure 12-6 and displaying it around your working area.

For some of the experiments in this manual, you are going to be using methanol, alkalines, and sodium borohydride. These will all get on your clothes and skin, and irritate or poison you if you are not careful, so do the sensible thing, and get yourself a lab coat. They are available in a variety

Figure 12-8 *Safety glasses—a must, and they make you look cool.*

of rather fetching styles, and in a variety of exciting colors—white! They also have the side-effect of making you look incredibly purposeful and professional in your experimentation, which is to be encouraged at all costs, as people will not then bother to distract you for chores or menial tasks, if you seem to be deep in the throes of making some discovery which could qualify you for the next Nobel prize.

The other thing which is a must-have, both for "the look" and safety is a pair of safety glasses. You only get one set of eyes, and if you flick something nasty in them, you could damage or impair your vision permanently, so make sure you wear them when working with chemicals, setting light to things, or cutting materials—where bits of material are liable to fly off and hurt you!

If you start getting serious about fuel cell experimentation, and decide to invest in large PEM electrolyzers, or regularly work with cylinders of hydrogen, or, if you are a school or university, you might like to consider investing in a gas detection system. Fuel Cell Store sell a hydrogen safety sensor, currently retailing at just over $100 (Part No. 570150), which could be a good investment as you get more serious about the hobby.

Above all, use your commonsense, and stay safe.

Chapter 13

Hydrogen Transport

Fuel cell cars

To appreciate the differences between fuel cell vehicles and other types of vehicles, we have to understand some of the energy transfers that are taking place. Ultimately, when we are talking about transport, we need to produce mechanical energy to turn the wheels of the car and make the thing go; however, although internal combustion engines can produce mechanical power directly, they only do so *efficiently* over a narrow power band, whereas electric motors produce a high torque over a wide power band. This is why car manufacturers are looking into hybrid vehicles to combine the strengths of both the power density that can be stored in fuels for the internal combustion engine, and the efficiency over a range of loads of the electric motor. See how battery vehicles and fuel cells fit into this energy conversion sequence in Figures 13-1 and 13-2.

Cars today tend to use four-stroke, internal combustion engines. These engines either burn petrol or diesel, but in a more limited capacity, some vehicles will burn LPG, biodiesel, or bioethanol.

While the effects of releasing carbon into the atmosphere can be mitigated by using different fuels such as biomass or biogas, it does not change the fact that the petrol engine, being a heat engine, is horribly inefficient, and converts not a lot of the chemical energy embodied in the fuel into actual delivered power.

In addition, there is another good reason for rejecting the internal combustion engine. Look at all the different byproducts that pollute our environment that come out of your exhaust pipe.

Furthermore, by burning fuel with air, a cocktail of nasty gases is released, which contains not only CO_2, but also NO_X and SO_X, as well as particulates.

In addition, internal combustion engines only really have a usable efficiency within a narrow power band. This means that we need a gearbox which adds further weight to the vehicle, and a

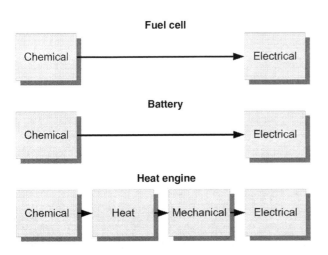

Figure 13-1   *Production of electrical power.*

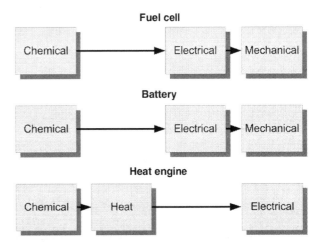

Figure 13-2   *Production of mechanical power.*

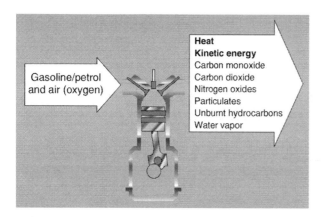

Figure 13-3 *The internal combustion engine produces all sorts of nasty emissions.*

Inside the figure:
Gasoline/petrol and air (oxygen)

Heat
Kinetic energy
Carbon monoxide
Carbon dioxide
Nitrogen oxides
Particulates
Unburnt hydrocarbons
Water vapor

clutch to engage and disengage the gearbox from the engine. All in all, this is not very satisfactory.

One criticism that people often have of fuel cell vehicles is that there is an argument for using fossil fuels to produce hydrogen. You, in effect, create a "long tailpipe" with all your vehicle's emissions coming from a centrally located, large tailpipe—the power station's chimney! This is a compelling reason why we need to look to clean, green sources of energy, in order to make sure that our future transport options have a reduced environmental impact.

Electric motors fit a vehicle's requirements much more closely. At low rpm, they produce the most torque—this is when a vehicle requires the extra energy to get it started. In addition to this, regenerative braking can be used, where kinetic energy from the car's forward motion can be recovered by using the motor as a brake, generating energy in the process which can be used to charge batteries and produce hydrogen.

The limitation for electric vehicles has always been the battery technology. Batteries are heavy, storing not a lot of energy for their weight. By contrast, hydrocarbon fuels provide a lot of energy for their weight, and hydrogen produces a fantastic amount of energy for its weight.

In addition to this, users have always favored convenience—a battery car takes a long time

to "charge up," and then can only run for a short amount of time.

Hydrogen sidesteps these problems with battery technology—a car can be fueled in around five minutes, and then the vast amount of power stored in the lighter-than-air hydrogen gas is converted to electrical energy, which in turn drives a conventional electric motor.

This gives us the best of both worlds—an energy-dense power source, with an efficient producer of kinetic energy.

There are already a large array of hydrogen fuel cell concept cars. A quick browse of any manufacturer's website will yield their particular implementation of fuel cell technology; however, most of these are based on the same fuel cell technologies from only a handful of suppliers. We can take a look at the "fuel cell cars gallery" at the end of this chapter for more details.

Fuel cell vehicle technology

When we look at a fuel cell vehicle, we notice that there are a number of significant differences. First and most apparent is under the hood—no internal combustion engine, just a fuel cell and electric motor!

Figure 13-4 *Under the hood of a fuel cell car.*

Figure 13-5   *Hydrogen storage tanks under the seat of a car.*

Secondly, a little more complexity needs to go into the engineering of the fuel tanks—rather than a fairly simple metal can, more complex arrangements need to be made for storing hydrogen. We explored some of these in the earlier chapter on storing hydrogen.

Thirdly, when we come to fill the car, rather than using a simple nozzle, we have a more complicated coupling that allows transfer of the hydrogen under pressure.

Figure 13-6   *A hydrogen coupling on a car.*

Iceland, which has committed itself to a hydrogen economy, opened the world's first hydrogen filling station in Reykjavik in April 2003. Hydrogen is produced onsite by electrolysis. The station provides hydrogen for a fleet of hydrogen-powered buses. Interestingly, there is no roof to this fueling station—this has been done to prevent the potentially explosive build-up of hydrogen which could accumulate.

A number of vehicle manufacturers have announced their intentions to develop, or currently have hydrogen fuel cell powered vehicles. That list includes: The Ford Motor Group, which incorporates Volvo, Mazda, and Jaguar; General Motors (responsible for producing Vauxhall cars in the UK); Honda; BMW; Nissan; and Hyundai.

Nearly all of the major car manufacturers are now showing demonstration models of fuel cell cars at world motor shows. While there are no fuel cell cars available to the general public at this time, the technology is certainly reaching maturity— manufacturers predict that we can expect to see fuel cell models launched around 2010 for general sale. However, one of the main barriers to widespread adoption is the lack of hydrogen filling stations, which will be an obvious concern for consumers.

At its North American headquarters in Torrance, California, Honda are exhibiting their "Home Energy Station," a package that promises to offer combined heat and power for the home, as well as providing a source of clean hydrogen for powering Honda fuel cell vehicles.

One of the main barriers to innovation is the cost of the exotic materials required to manufacture the fuel cells. The platinum used for catalysts is expensive, so current research is focusing around trying to eliminate, or at least reduce, the amount of platinum inherent in the design of fuel cells. There are already promising results—as recently as 2002, fuel cells had a platinum content of £600 per kW output; however, we now have cells containing around £20 per kW output. Further advances will doubtless be made.

Figure 13-7 *A specially constructed hydrogen filling station.*

For long-distance, low-occupancy journeys, motorbikes present a more efficient solution for personal transport than cars. What is the sense in transporting over a ton of vehicle, to move a person who weighs 80 kg?

A number of manufacturers have launched small bikes and mopeds powered by fuel cells. Particularly promising is the ENV fuel cell bike, which offers a stylish alternative to conventional petrol bikes.

In Europe, there are a number of countries trying out fuel cell buses as a clean alternative for public transport. While they have not been widely adopted, the technology is under test and development, and shows much promise for the future.

Figure 13-8 *ENV fuel cell bike.*

Figure 13-9 *A London bus—powered by hydrogen. Image courtesy Ballard.*

In this experiment, we will be building a simple fuel cell car, which you can use as a base for experimentation to investigate the science of fuel cell vehicles. This car is of a very simple design, and functional rather than aesthetic. There are competitions for fuel cell vehicles such as this, and there is a lot of potential for flexibility in the design of fuel cell cars. Once you have built this simple model, you should have a grasp of the basic principles of fuel cell vehicles. Use this project as a springboard for experimentation, and try and enhance the performance of this simple model—make it more lightweight, add bearings, reduce friction, improve balance, improve contact with the road–the list of engineering topics where this simple vehicle can take you is endless.

You Will Need

- Fuel cell car competition kit (Fuel Cell Store Part No: 7110303) containing:

- Reversible fuel cell

- Small motor with wires

- Motor mount

- 4 small screws for motor mount

- 1 small screw to secure motor

- Tubing

- 4 wheel hubs

- 4 O-ring tires

- Generous selection of small gears

- 2 axles

- Corrugated plastic, correx or corroflute

- Hydrogen and oxygen 30-ml cylinders (Fuel Cell Store Part No: 7110307)

Tools

- Box of matches

- Philips screwdriver

- Craft knife

Hint

You can purchase a kit (Figure 13-10) from Fuel Cell Store, which contains all the fuel cell and mechanical components needed to produce a simple fuel cell car. The kit comes with wheels, tires, axles, gears—everything except the chassis and storage of hydrogen—so this is a good way to get started.

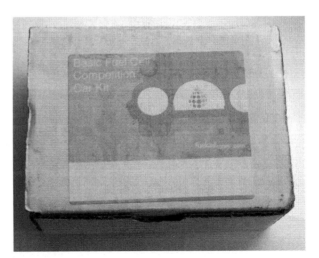

Figure 13-10 *Kit of components for the simple fuel cell car.*

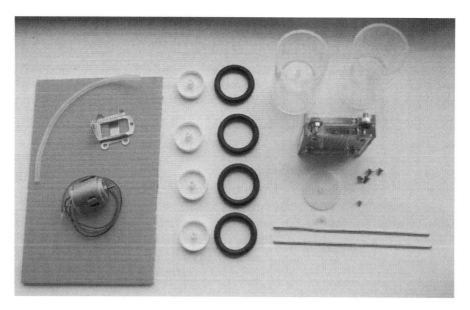

Figure 13-11   *The components for the simple fuel cell car.*

Once everything is out of the box, and you lay it out, it should look something like Figure 13-11. I tell you to get corrugated plastic for a reason. It is perfectly possible to buy corrugated card or foamboard, or thin plastic, or any other manner of materials from which to make your fuel cell car base. However, by its nature, corrugated plastic is very lightweight, yet stiff and rigid. It also has the useful property that the plastic tubes created by the corrugations provide a *very* convenient means of anchoring your axles, and remove all the complexity of bearings and axle mounting. As you experiment, and decide to work with different materials—without the corrugations—small lengths of drinking straw can be used instead to hold the axles in place.

The first step is to take the plastic wheel hubs (there should be four of them) and the rubber O-rings out of the box. It is a simple matter of pushing the O-rings onto the plastic hubs to form the finished wheels. (Take a peek at Figure 13-12.)

The next stage is to assemble the motor into its metal mounting cage. The motor has a circular protrusion at one end, which supports the rotating shaft. The metal cage has a circle to match this at one end. The motor should be mounted with the shaft pointing through that hole, as shown in Figure 13-13.

Once you have done this, look at the back of the motor. You will see that the bottom of the metal cage protrudes past the back of the motor. You will

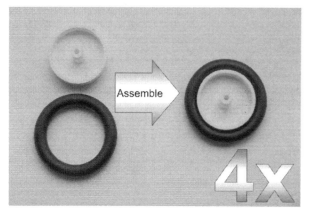

Figure 13-12   *Assemble the wheel hubs and O-ring tires.*

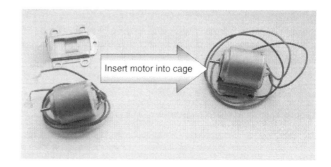

Figure 13-13   *Insert the motor in the cage.*

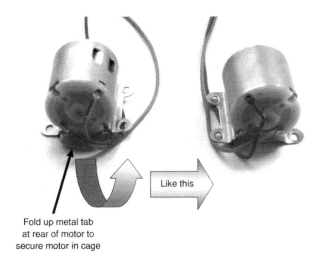

Fold up metal tab
at rear of motor to
secure motor in cage

Figure 13-14 *Fold the tab to secure the motor.*

also notice on the plastic backing of the motor that there are two small holes. Firstly, you should be able to gently bend this piece of metal over the back of the motor, so that the small hole in the center aligns with the small hole on the back of the motor.

It is then simply a matter of taking the smallest screw supplied in the box, and anchoring the motor to the cage by securing it, as shown in Figure 13-15.

Now, we come to manufacturing the chassis. For this, you will need a small piece of corrugated plastic which is 115 mm by 175 mm. One of the things that you need to ensure is that the "corrugations," or lines in the plastic, run across the short side of the plastic. This is essential as, as mentioned before, we will be using these corrugations to mount our axles.

Refer to the cutting diagram for the next step. The diagram is printed to scale, so you can either transfer the measurements manually onto your

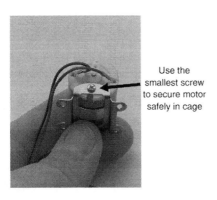

Use the
smallest screw
to secure motor
safely in cage

Figure 13-15 *A screw secures the motor in the cage.*

sheet of plastic, using a marker rule and set-square, or you can photocopy this page and use it to mark directly onto the plastic.

Now mark out the plastic, and it should look like Figure 13-17.

The best way to cut corrugated plastic is to use a sharp craft knife and a cutting board. Because some of the cuts are *internal*, a pair of scissors will not really be suitable. You will see that rectangles of material will need to be removed from the board—this is to provide space for our wheels. Do this first as this is the easiest step. Now, see the little rectangle which is off-center. Cut through both layers of plastic, and remove this piece of plastic from the board—this is waste and can be discarded.

Now, with the "H" shape in the centre, we need to be careful. We are not removing this piece of plastic, but we do need to bend it in order to support our fuel cell. Cut through both layers of the "H" marked in bold.

Now, one of the properties of corrugated plastic is that if we want to make nice, clean folds, we can cut through one layer of plastic, but leave the other. So, along the dotted lines, carefully use your craft knife to cut through one side of the plastic but not the other.

Once all the excess plastic has been removed, the final blank chassis should look like Figure 13-19.

The first step is to screw the motor and its mounting to the board. If you transferred all the measurements, you should see four small crosses. These are the center-points where we will locate the screws to hold our motor mount down. The small self-tapping screws bit readily into the plastic, and do not need any predrilling or center punching. You need to locate the motor, so that the shaft is hovering over the slot which we cut all the way through, as shown in Figure 13-20.

Now, we need to select some appropriate gearing for our vehicle. A compromise will have to be made here. Our motor provides a certain amount of "torque", that is to say, "turning force," and with a fixed amount of power, it will run at a certain speed.

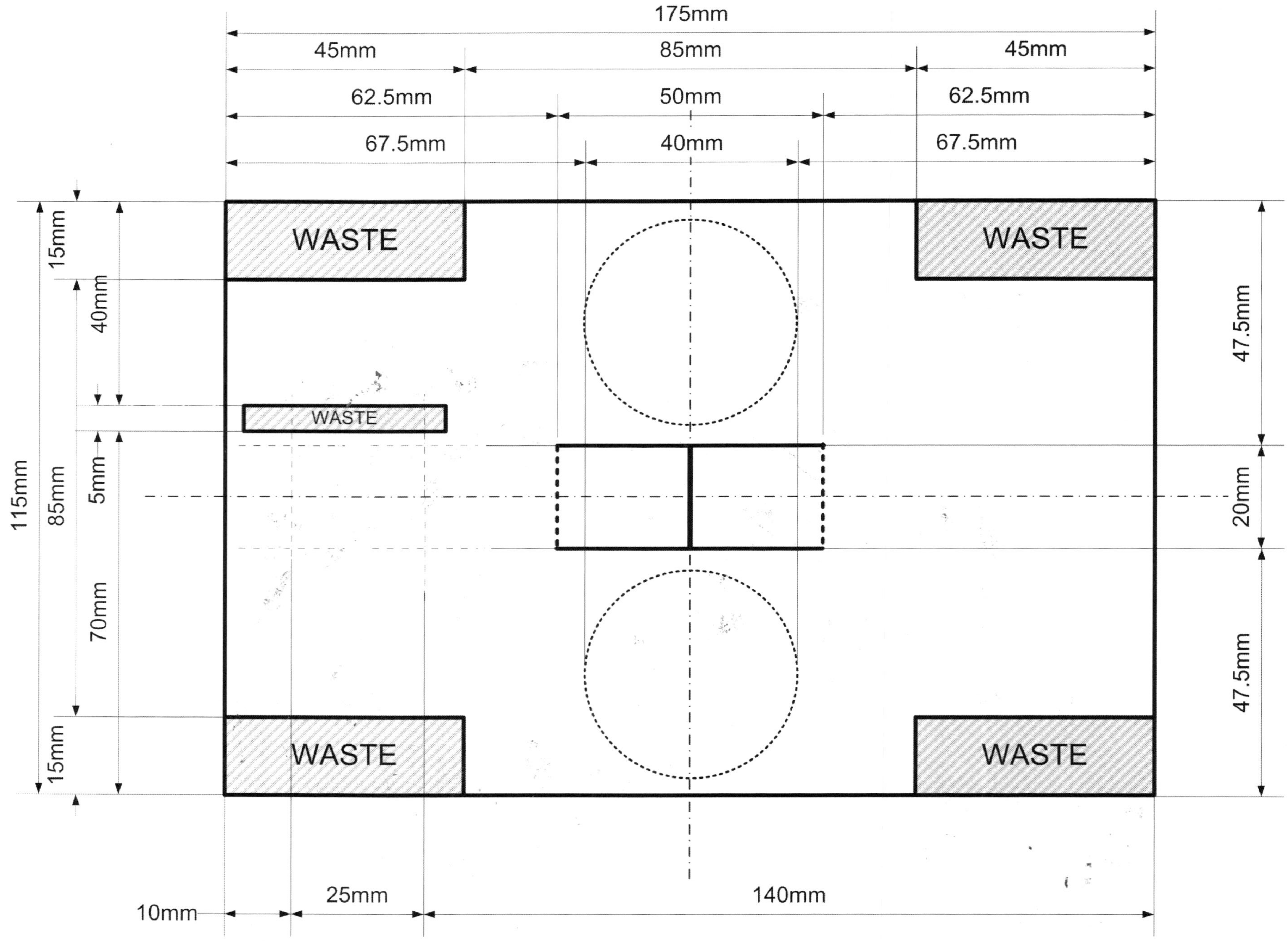

Figure 13-16 Cutting diagram for the chassis.

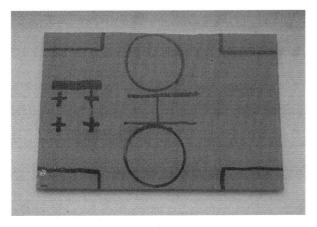

Figure 13-17 *The plastic chassis marked out ready for cutting.*

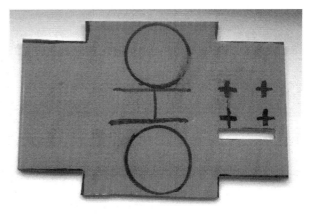

Figure 13-19 *The plastic chassis cut-out.*

We want to optimize the relationship between speed and torque when transferring the power of the motor to the wheels on the ground. We can use gear ratios to take the speed of the motor, and *reduce* the speed, but improve the torque. This gives our vehicle *pulling power*. However, remember, we trade torque for speed. Make the vehicle too powerful, and it will only go very slowly. Make the vehicle too fast, and it may not have sufficient power to climb even the slightest incline, or navigate a rough surface. In addition, we can use compound gear ratios. We will not be using them in this model; however, it is something you should be aware of. We can use gears to turn gears in effect, reducing or increasing the ratio each time.

If a shaft spins at a set speed and it has a gear with 10 teeth on the end, this gear with ten teeth

then engages with a gear with 60 teeth—the output speed will be one sixth of the original speed, but the turning force will be six times higher. We can see how speed and power will be affected simply by counting the teeth on the gears, and comparing the ratios. We can see some different arrangements of gears in Figure 13-21.

Your first fuel cell car is going to have simple gearing—a small spur gear driving a larger gear which will drive the axle which powers the wheels. Select the smallest gear from the assortment—the

Figure 13-20 *Motor mounted with shaft above gear-slot.*

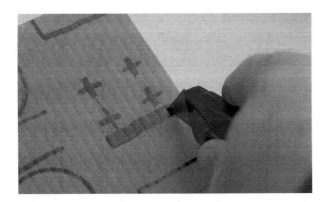

Figure 13-18 *Use a sharp craft knife to cut through both layers of the plastic.*

Figure 13-21   *Combine the gears to make different ratios.*

one with the least teeth, and push it onto the shaft of the motor, as shown in Figure 13-22.

Now, we are going to install the fuel cell—remember, we cut the board on one side of the "H". . . well, fold the flaps up to provide support for the fuel cell, as shown in Figure 13-23.

Now, you can either use a little sticky tape, or even some double-sided tape on the flaps, in order to fix your fuel cell in place, like the image in Figure 13-24. The hole is the correct size for you to *wedge* the fuel cell into place, with the hole gripping around the bottom of the cell providing additional support.

Next, we are going to mount the rear axle and gear wheel. Take one of the medium-sized single gears, and hold it in the rectangular slot. The correct-sized gear will allow the hole through the gear to align with one of the rectangular channels in the corrugated plastic, as shown in Figure 13-25.

You may find that the axle supplied is a little too tight for the gear, so in this instance we can use a small drill to enlarge the hole to a suitable size for the axle.

Next, push the axle down the channel, through the hole in the gear, and onto the other side of the plastic, and out the other side. It should look something like Figure 13-27.

Now for attaching the wheels. First start with the front wheels. Again, you may find that you need to drill out the wheels slightly. Another labor-saving way to make things fit is to heat the plastic to soften it, allowing the axle to slide in. The way to do this is to heat the axle till it is hot, and push it into the plastic wheel. This will cool to make a very firm fit. Make sure you heat the axle and push

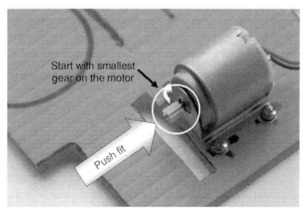

Figure 13-22   *The smallest gear should be a push-fit on the motor shaft.*

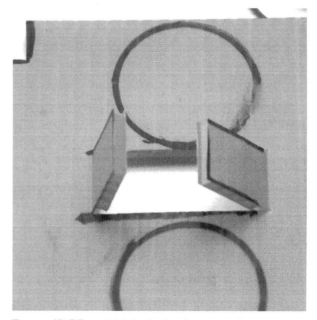

Figure 13-23   *Fold the fuel cell support flaps up.*

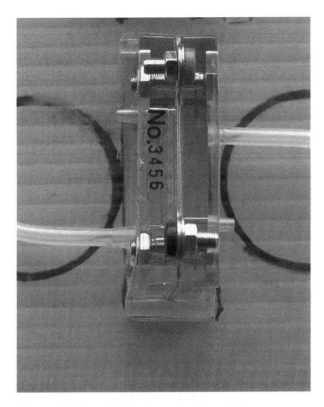

Figure 13-24   *Fuel cell mounted in position.*

it into the wheel, not vice versa, for a good result. See Figure 13-28 for comparison.

Once all of the wheels are attached, connect the motor to the terminals of the fuel cell, and using short lengths of hosing, connect the fuel cell to the

Figure 13-25   *Align the axle hole with the corrugated plastic holes.*

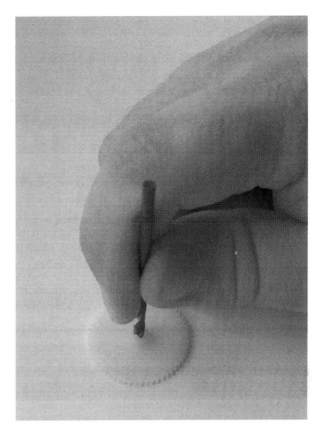

Figure 13-26   *Use a small drill to enlarge the axle hole.*

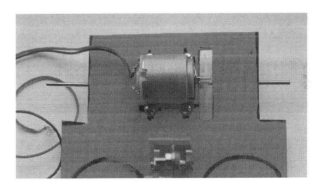

Figure 13-27   *Axle and large gear wheel installed.*

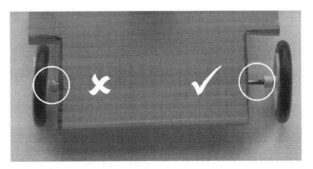

Figure 13-28   *Heat the axle not the wheel.*

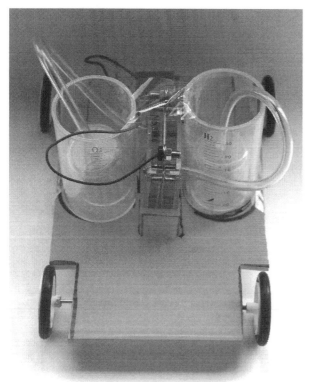

Figure 13-29   *Simple fuel cell car complete (front view).*

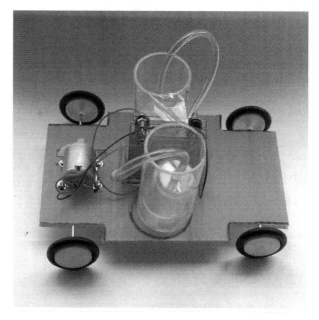

Figure 13-30   *Simple fuel cell car complete (side view).*

respective hydrogen and oxygen tanks, being sure to put a small plug in any unused ports in the fuel cell. You will find that a small piece of double-sided tape or sticky tape will be sufficient to anchor the storage tanks in their correct position.

And there you have it! Our completed fuel cell car! See Figures 13-29 and 13-30 for how the finished model should look.

You will now need to follow the procedure for charging the fuel cell with distilled water, and filling the tanks, then creating hydrogen and oxygen using electrolysis. You will then generate sufficient hydrogen to run the car for a small distance.

Project 34: Building an "Intelligent" Fuel Cell Car

You Will Need

- Intelligent fuel cell lab (Fuel Cell Store Part No. 550190)

If you just want a fuel cell car that you can play and experiment with out of the box, then the intelligent fuel cell car is just the ticket! It provides a ready-made chassis, and it has a clever differential

gearbox and steering system, which allows the car to change direction when it bumps into obstacles. Take some foamboard, make a small enclosure with walls 4 in-high, and your intelligent fuel cell car will happily trundle around crashing into walls, and then reversing in a different direction.

Get the components for the car out of the box (Figure 13-31a), and they should look like Figure 13-31b.

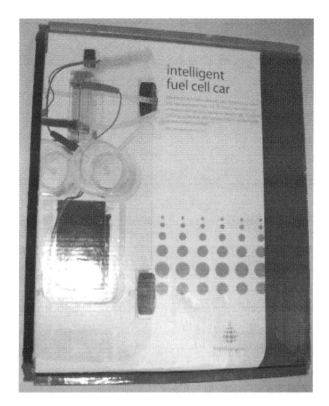

Figure 13-31a   *Intelligent fuel cell car in the box.*

The first step is to press the wheels onto the stubs protruding from the plastic chassis—your car will look like Figure 13-33. None of the wheels are "driven"— the action takes place in the little black box in the center of the car, which provides all the power.

Next, connect up your tanks to the fuel cell— you should be used to this procedure by now!

The chassis comes with a special area to clip the tanks and the fuel cell to it. Install them now.

You're likely to have a little bit of distilled water splashing about, so install the cover over the motor mechanism to prevent it from getting wet, as shown in Figure 13-35.

Now, a handy plug is provided to charge the car from the supplied battery pack. When you are using the PEM fuel cell in the electrolyzer mode, you will need to connect the leads from the plug to the fuel cell, as shown in Figure 13-36.

However, when you want the car to "go," you will need to connect the motor unit to the fuel cell, as shown in Figure 13-37.

And that's everything complete! Ladies and gentlemen . . . The intelligent fuel cell car is in Figure 13-38.

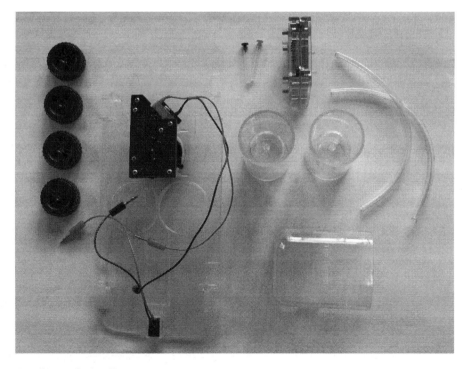

Figure 13-31b   *Intelligent fuel cell car components.*

Figure 13-32   *Wheels installed.*

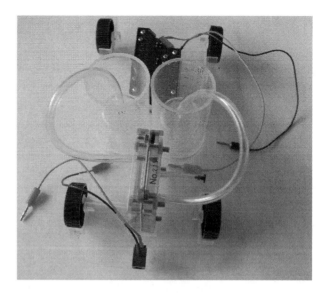

Figure 13-34   *Fuel cells and tanks installed.*

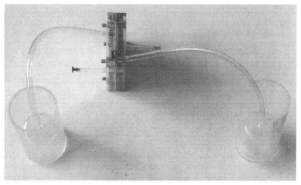

Figure 13-33   *Fuel cells and tanks plumbing.*

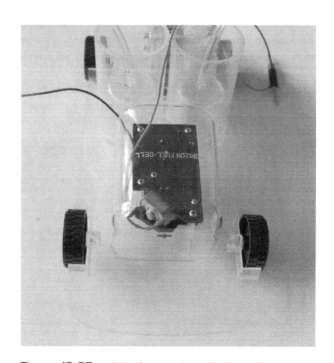

Figure 13-35   *Splash cover installed over the motor.*

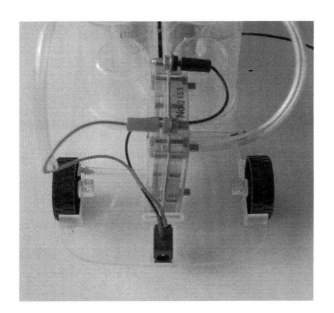

Figure 13-36   *Fuel cell connected to the battery slot.*

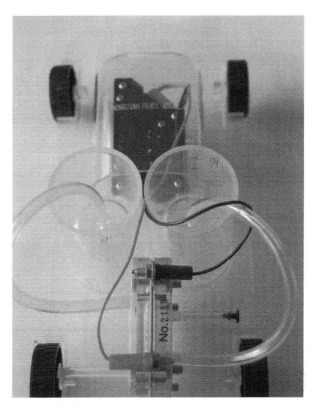

Figure 13-37   *Fuel cell connected to the motor.*

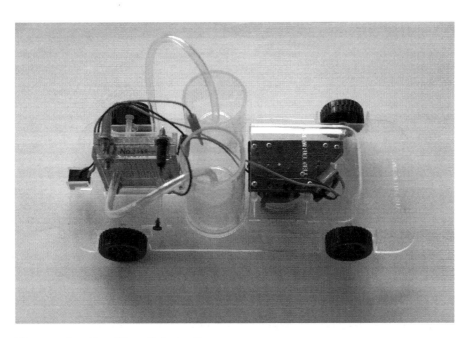

Figure 13-38   *The complete "intelligent" fuel cell car.*

Project 35: Building a Hydride-Powered Fuel Cell Vehicle

- Simple or intelligent fuel cell car (see above)
- H-Gen hydrogen generator
- Sodium borohydride

Another approach that has been investigated for fuel cell vehicles is to use hydrides for storing hydrogen. This has the advantage that hydrides are easier to store than hydrogen gas; however, there is an associated weight penalty.

It has been speculated that vehicles could be "fueled" with a hydride, which is reformed onboard the vehicle using a catalyst. Once the catalyst exhausts the supply of hydrogen from the hydride, the spent fuel is returned to a holding tank. This spent fuel could then be offloaded from the vehicle at a filling station, and regenerated chemically to its previous state, ready for the next customer. The hydrogen would have to be separated from the liquid hydride, then any heat produced from the reforming process would be extracted using a heat exchanger, with the hydrogen being fed to a fuel cell to produce electrical power, which would then power electric motors. A schematic of this process is shown in Figure 13-39.

Remember the H-Gen that we used in the chapter on storing hydrogen (Chapter 4)? Well, let's try powering our fuel cell vehicle from hydrides!

Produce hydrogen and oxygen as you normally would by electrolysis. Save the oxygen in the storage tank but *open* the hydrogen tank, and allow it to vent so there is no hydrogen gas. Now, put a little sodium borohydride in the distilled water, and drop the catalyst pill from the H-Gen

into the water. As hydrogen is produced, hold the storage part of the gas cylinder over the fizzing pill to capture hydrogen. Wait as a little hydrogen builds up under the cap, and then press it firmly home.

You should be able to power your fuel cell vehicle from the hydrogen, which is being released from the hydride.

Hint

The limiting factor is now the fuel cell's supply of oxygen. As we have seen, fuel cells can function on air, but the performance is likely to be fairly poor, and your fuel cell may not produce enough power to move your vehicle.

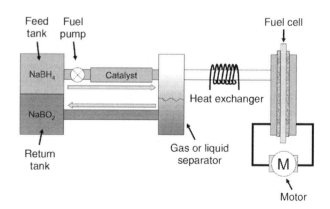

Figure 13-39 *Schematic of hydride vehicle.*

Project 36: Building a Radio-Controlled Fuel Cell Vehicle

You Will Need

- A radio-controlled car
- PEM fuel cell (the type used in the last two experiments is suitable)
- Hydrogen storage tank
- Oxygen storage tank
- DC-DC convertor
- Short lengths of wire
- Small piece of corrugated plastic, correx, or corroflute
- Double-sided tape

Tools

- Electrician's screwdriver

Hint

When choosing a radio-controlled vehicle, you need to look for something small that will run on 4.5 V—larger radio-controlled vehicles require a lot of juice, and have large battery packs—our small fuel cell will wimp out at the sight of them, so look for something a little more modest that runs from either AA or AAA batteries.

In this project, we are going to convert a radio-controlled (RC) car to run on fuel cell power! We are going to take an off-the-shelf car and soup it up, to run it from clean, green hydrogen power.

We can't simply connect our fuel cell directly. The reason for this is that the motors we used in the previous two projects were small low-voltage motors. They are quite happy to run from the low voltage produced by a single-cell fuel cell; however, the needs of even a small RC car are a little more demanding. We are going to boost the voltage using a DC-DC converter. Remember the characteristics of fuel cells that we discovered? They tend to produce a fair amount of current, but the cell voltage is the limitation. Well, a DC-DC converter helps us get around this limitation by sacrificing current for voltage. We are going to use a small DC-DC convertor to boost the single-cell power of our fuel cell up to a voltage suitable for powering a little RC car.

When you have selected your vehicle, assemble all your components, and you should have something that looks a little like Figure 13-40.

Now, it must feel like Christmas come early—open the box of your RC car, and take out the car. The handset controller will be powered by ordinary batteries, so now would be a good time to fit some.

The battery compartment of your RC car will most likely have a screw to secure it, so take a small screwdriver, remove the screw and open up the battery compartment of your RC car.

I recommend that you use solid wire for attachment rather than flexible wire—there are a couple of reasons for this—it makes the next step easier, and furthermore, it allows you to "bend your wire to shape," making sure that it doesn't snag the wheels of your RC vehicle.

Figure 13-40   *Components for RC fuel cell car.*

You need to locate the two single terminals. One will be a spring (this is the negative) while the other will be a flat plate (the positive)—the polarity and way that the batteries are inserted should be clearly labeled inside the compartment. Ignore the double terminals—these are just used to make a contact path between the batteries. The dotted white line shows the path that the electricity would normally flow between the batteries.

We are going to be "piggy backing" onto the battery terminals—if you are only going to use the car for this project, then you might consider a permanent method of attachment. As a rule, it will be hard to solder wires onto your battery contacts if they are made of stainless steel, so you might look at conductive glues which do a similar job to solder, but without the heat (and they take a little longer to cure).

The method which I have found is best for attaching wires is to use a very fine electrician's screwdriver and lever the battery terminal forward

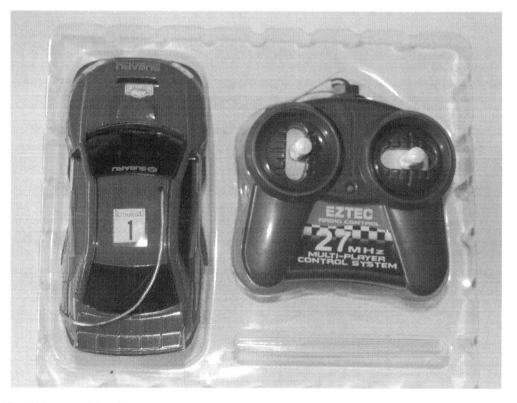

Figure 13-41   *RC car and handset.*

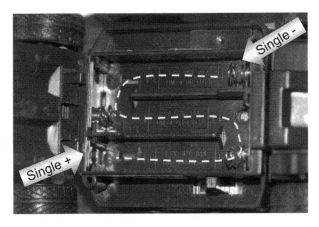

Figure 13-42 *RC car battery compartment—terminals labeled.*

Hint

You could always check for a good contact now using a wall-wart power supply (but make sure it's set to the right polarity, or else you will fry the electronics of your vehicle.) It's probably easier to do this now as a fault-finding exercise than to wait until everything is assembled.

a little, then try and slide the solid core wire in the space between the terminal and the plastic of the car body. This will be a tight fit, and should be sufficient to reliably hold the wire in place.

Once you have made a good electrical contact, shut and screw back the battery compartment, trapping your wire in the space between the battery flap and door. (If the worst comes to the worst, and you can't find a way out for your wires, you could always use a small needle file to file a small groove to allow your wires to pass through.)

Now flip the car over, and you should end up with something like Figure 13-44.

Now, to power our little automotive beastie, we are going to need the previously mentioned DC-DC converter. If you want to take the hassle out of things, you can just buy a little board off-the-shelf (the Fuel Cell Store do one—part number listed above). However, if you're a dab hand with the soldering iron, look at Figure 13-45 to see how you can make your own.

Figure 13-43 *Underside of the RC car with wires leading to the battery compartment.*

Figure 13-44 *RC car with power wires.*

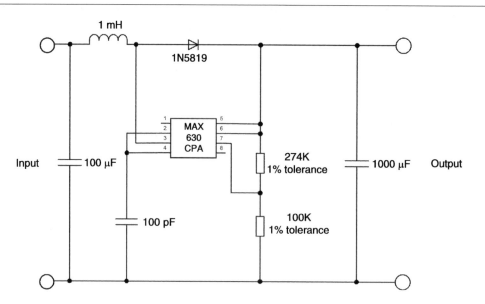

Figure 13-45 *DC-DC converter board schematic.*

If you have some skill with hobby electronics, a DC-DC converter board is very easy to make, as an integrated circuit removes many discrete components.

You will see, either on the schematic or the PCB, depending on what option you have selected, that there is very clearly an IN and an OUT. The DC-DC converter is polarity sensitive, so make sure that you get things the right way round! It should seem fairly obvious that the OUT end connects to the wires coming from under our RC car, while the IN end leads to the fuel cell. In the shop bought board, you just need to take an electrician's screwdriver, and secure the stripped ends of the wire in their terminal blocks.

Car roofs tend to be curvy streamlined affairs—great for aerodynamics in the real world at high speed, but not a good surface for mounting fuel cells! We, however, are going to take a small sheet of corrugated plastic—lightweight and strong—and use it at a platform for affixing our components. First of all, we are going to cover it in double-sided tape, on both sides, in order to provide a sticky surface to easily mount our components. The beauty of using double-sided tape is that a strong tape should provide sufficient mechanical

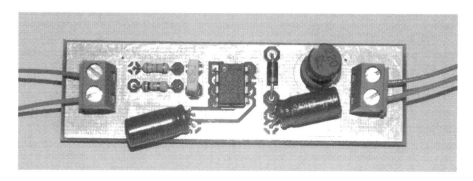

Figure 13-46 *DC-DC converter board from the Fuel Cell Store.*

Figure 13-47 *Cover the plastic platform with double-sided tape.*

support for all the components while you experiment, but is weak enough that should you wish to use the components (and the RC car) again, you can easily disassemble them with a little care—and they won't be damaged.

Now we come to the powerhouse for our teeny motor! Remove the backing from the double-sided tape, and position your hydrogen tank, oxygen tank, and fuel cell on the sticky surface. This will hold everything in place. Now, remove the backing

from the double-sided tape on the opposite site of the plastic, and stick the assembly on the roof of your car. Now, taking the tubes, connect up your fuel cell to the tanks—you should know the drill by now!

Once you have connected up the DC-DC inverter, fuel cell, and car battery terminals, you can think about mounting the equipment on your vehicle. Double-sided tape is relatively easy to use, and generally won't damage the vehicle underneath if you want to change things around. For mounting the DC-DC convertor, you could employ a similar approach, or alternatively you could mount it using Blu-Tack.

Make sure that the wires are kept out of the way from any wheels so that they do not snag.

The finished vehicle is shown in Figures 13-49 and 13-50—it's not going to win any awards for automotive design brilliance, but it is certainly one of the greenest cars on the block.

If you wanted to get more advanced, you could consider removing the vehicle components from its enclosure and constructing a new body, but this is purely an aesthetic issue—the function of the car will remain the same! If you do decide to go down this route, you will be confronted by similar issues

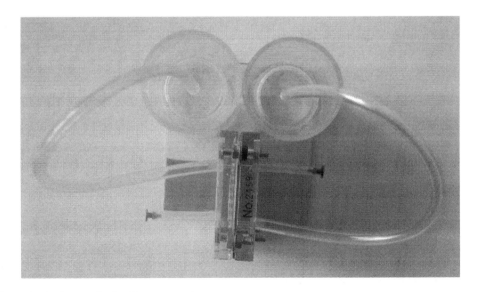

Figure 13-48 *The hydrogen fuel cell power unit.*

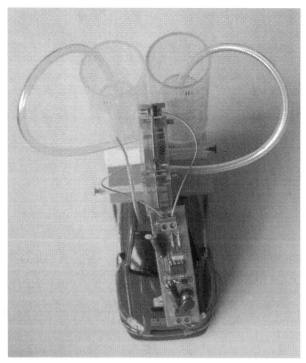

Figure 13-49 *The completed RC fuel cell car (top view).*

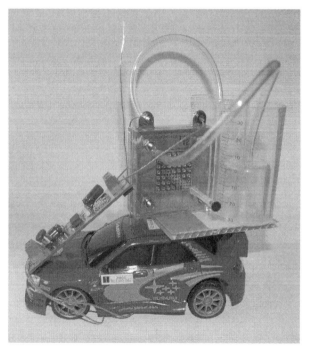

Figure 13-50 *The completed RC fuel cell car (side view).*

to the major automotive manufacturers—where to mount the hydrogen storage tanks, which are somewhat big and bulky.

Once your vehicle is complete, follow the procedures in the PEM fuel cell chapter (Chapter 7) for charging your fuel cell with distilled water, producing hydrogen by electrolysis, and then connect the DC-DC convertor back for some rip-roaring racing action!

You could consider building an obstacle course for your car and seeing how well it performs. Remember, the challenge isn't just to stay on the road, but also to stop distilled water from slopping out of the rooftop storage tanks, as these are providing the pressure for the hydrogen (and oxygen) underneath.

You will find that your vehicle performs much better on smooth surfaces. Small RC cars do not always perform very well on carpeted surfaces. Also, if you stick to tiled and laminated surfaces, it is easier to wipe up any distilled water that is spilt in a major collision!

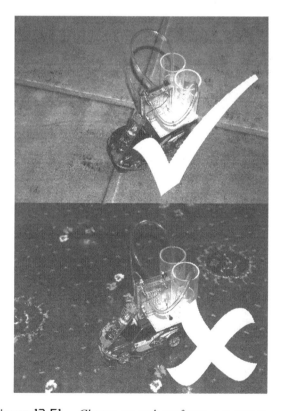

Figure 13-51 *Choose smooth surfaces, not carpet.*

Fuel cell planes

And now, we move swiftly on from fuel cells on the road to fuel cells in the sky—we are going to have a look at some of the different ways that hydrogen could be used in the future to transport us from place to place.

At the moment, the aviation industry is growing at an unprecendented rate. It's now possible to fly cheaply and comfortably—a dream that only 50 years ago was out of reach for many people. However, this boom in aviation and travel comes at a cost—a carbon cost. Airplanes produce vast quantities of carbon dioxide, as their jet engines ooze out the emissions from burning hydrocarbon kerosene. If we are to look to a future beyond oil, we need to look at other alternatives that can parachute us into the 21st century.

Using hydrogen in the air isn't a new phenomenon. The Russian airplane company, Tupolev, successfully flew a tri-jet airliner in 1988 in the former Soviet Union, where one of the engines was powered by hydrogen that was stored onboard cryogenically. The aircraft was a Tu154 airliner that had been specially adapted.

Hydrogen is a fuel that is well suited to use in aircraft due to its tremendous energy—by weight; however, as a result of its poor energy density by volume, it is apparent that future aircraft fueled by hydrogen will require tanks that are four times bigger than those employed on current aircraft.

NASA introduced fuel cells to flying with its Helios prototype unmanned plane. The plane could provide a substitute for expensive satellites, by orbiting the earth, unmanned, powered by clean, green energy from the sun during the day, its large wing structure being covered in photovoltaic cells, while producing hydrogen from spare power. Then, during the night when the sun is hidden away, the reserves of hydrogen generated during the day are converted to electrical energy to drive the plane's propellers and maintain flight during the night.

Figure 13-52   *NASA Helios prototype. Image courtesy NASA.*

A little later in this chapter, we are going to look at lighter-than-air travel, but next, let's blast off into outer space!

If you want to read some more about hydrogen-powered planes, the following are a good selection of Web links:

www.flug-revue.rotor.com/FRheft/FRH9809/FR9809k.htm

www.bl.uk/collections/patents/greenaircraft.html

www.nasa.gov/centers/dryden/history/pastprojects/Erast/helios.html

Project 37: Space Travel with Hydrogen! Building a Hydrogen Rocket

You Will Need

- Manganese (IV) oxide
- Zinc scraps
- Hydrochloric acid (4.0 mol l^{-1})
- Hydrogen peroxide (3%)
- 2 × boiling tubes
- 2 × rubber bungs

    or

- PEM electrolyzer
- Power source

    and

- Gloves
- Safety specs
- Microbore tubing
- Cork lid
- Paperclip
- Plastic pastettes

Tools

- Tesla coil
- Awl

You've seen how metal scraps and acid can be used to produce hydrogen—well, in this experiment, we are also going to use manganese oxide and hydrogen peroxide to produce oxygen. Alternatively, if you haven't got that sort of stuff hanging around, you can produce hydrogen and oxygen using your PEM electrolyzer.

Hint

If you aren't sure where to get a Tesla coil from, talk to a good science lab technician at a school or college, or check out Bob Iannini's *Electronic Gadgets for the Evil Genius*, McGraw-Hill, 2004, which comes with some fab ideas for high-voltage electric-spark-making mayhem!

Making a suitable bung

You will need to take a rubber bung that fits your boiling tubes, and drill a small hole in it using an awl or a small drill, through which you can insert some microbore nylon tubing. You will use this bung to help you collect the gases generated by the chemical reactions. Alternatively, if you are using your PEM electrolyzer, you will need some microbore tube "epoxied" into the end of the standard rubber tubing, in order to squeeze the tube into the opening of the pastette.

Producing hydrogen

Take a clean boiling tube, and drop in a couple of small pieces of zinc and a little 4.0 mol l^{-1} of hydrochloric acid, so that there is about two inches of fluid in the tube. The hydrogen gas will bubble. Using a similar bung to that described above, you will be able to displace some of the water in the pastette.

Producing oxygen

On the tip of the spatula, take a small quantity of manganese (IV) oxide and place it in a boiling tube. Now add 3% hydrogen peroxide. You should have about 2 inches or so of fluid in the bottom of the tube. Now quickly stopper the tube with a bung—it will have a small length of tube attached. Insert the tube into the pastette.

Building a launch pad

You need a small rod to launch your hydrogen rocket, so take a small paperclip, straighten it out, and stick it into a large cork lid from a jar. The pipette can then slide onto the rod.

Preparing for launch

Using the two boiling tubes full of hydrogen and oxygen, fill a small pastette half with hydrogen and half with oxygen. Leave about an inch of water in the bottom. This helps to seal the gases inside the tube, and furthermore provides a reaction mass to produce some meaningful thrust.

Launching the hydrogen-fueled rocket

Hold the sparking end of the Tesla coil near to the base of the pastette where the gas meets the water. After several sparks, you will hear a loud pop, and your pastette will fly off the end of the paperclip, soaring into the sunset . . . well, not quite! At least over the other side of the room!

If all that seems like a little bit too much hard work, Estes, the model rocket company, now sell a hydrogen rocket kit, where hydrogen produced by the process of electrolysis is used to launch a small rocket from a launch pad. The solution comes ready-made—with this book, you now understand the science!

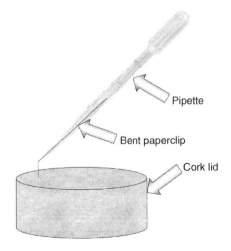

Figure 13-53 *Pipette hydrogen rocket.*

If you want to go beyond hydrogen, and explore more aspects of amateur rocketry, check out my book *50 Model Rocket Projects for the Evil Genius*, McGraw-Hill, 2006.

Figure 13-54   *Estes hydrogen rocket kit.*

Project 38: Air Travel with Hydrogen! Building a Hydrogen-fueled Airship

You Will Need

- Fuel cell mini-PEM (Fuel Cell Store P/N:531907)
- Small plastic propeller
- Small, efficient motor (e.g., pager motor)
- Length of plastic tubing
- Sheet of Mylar

or

- Mylar helium balloon

Tools

- Household iron

As much as this experiment is about building an airship, it is also about learning as regards the density of hydrogen gas relative to air.

We can produce hydrogen, using any of the methods in this book, to fill a sealed envelope of lightweight material. You could use an off-the-shelf helium balloon, filled with hydrogen, or you could make your own envelope from Mylar which can be easily heat-sealed with a household iron. Take a tube from the balloon, and epoxy it to one of the mini fuel cells we encountered in Chapter 7. The power from this PEM fuel cell can be used to feed a small electric motor, which in turn can be used to turn a model toy propeller to create a moving (if somewhat erratic) blimp!

You will find that, as hydrogen is consumed to power your motor, the blimp will gradually lose height, so our blimp will only make short flights. But you can always recharge your blimp by

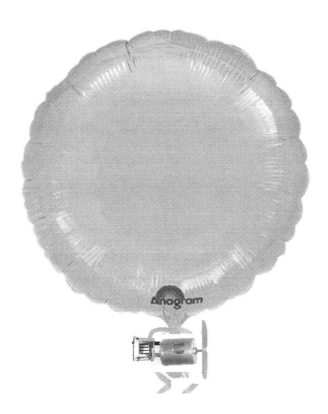

entry point of the pipe is well sealed to prevent the escape of hydrogen. A little dab of glue (or the ubiquitous duct tape) goes a long way to keeping things snug.

Remember, if you find that you are not getting sufficient *push* from your propeller, try reversing the polarity of the motor and see what effect that has.

The fuel cell, motor, and propeller can all be mounted on the *tail* of the balloon, using a few dabs of hot melt glue, or even tape.

Compensate for the buoyancy of the hydrogen by adding a little plasticine or modelling clay to the bottom of the balloon.

Figure 13-55   *A home filled, powered blimp from a standard helium balloon.*

making some more hydrogen with your PEM fuel cell, making sure to anchor the blimp down when filling it, so it doesn't go drifting off.

Helium balloons come with a handy hole and one-way valve for insertion of a nozzle; however, you will have to make sure, when fixing your hydrogen supply pipe for your fuel cell that the

Warning

Helium is generally used as the gas of choice for filling balloons rather than hydrogen as it is nonflammable; however, hydrogen has more lifting power, at the expense of being slightly more dangerous! If you are going to fly a hydrogen balloon, think very carefully about the areas you fly it in, and ensure that there are no sources of ignition present.

Project 39: Building a Radio-Controlled Airship

If you want to take the last experiment further, you might like to consider the possibility of building an airship that you can control and steer

from base control using radio control. You will be using hydrogen for its lifting power, and the actual motors will be powered by batteries in the radio control kits; however, you might like to consider how a DC-DC convertor could be used to power some of the lower-powered kits. Also, remember, these kits are designed to work with helium, as it is generally seen as safer (it doesn't explode). If you are going to use hydrogen in these blimp kits, think *very* carefully about where you are going to fly them, as you don't want to be causing fireballs, do you? Make sure you stay away from areas where there could be a naked flame, or other source of ignition.

www.johnjohn.co.uk/shop/html/model_ radiocontrol_blimp.html

www.hobbytron.com/mach3_rc_blimp.html

store.makezine.com/ProductDetails.asp?Product Code=MKBLIMPKIT

Chapter 14

More Fun with Hydrogen!

In this chapter, we are going to explore some different things that you can do with hydrogen—some of these projects are good for demonstration, showing and proving that hydrogen has genuine applications in the real world. These experiments are presented more as an illustration of what can be done than as practical objects to use every day; however, as the technology gains momentum, expect some of the ideas in this chapter to become mainstream.

Project 40: Powering a Radio Using Hydrogen

You Will Need

- 100 μF capacitor
- 100 pF capacitor
- 1mH inductor
- 1N5819 diode
- 274 K resistor (1% tolerance)
- 100 K resistor (1% tolerance)
- 100 μF capacitor
- USB socket
- MAX 630 CPA integrated circuit

Yet another application for our fabulous DC-DC converter featured in the radio-controlled fuel cell car project—why not use a fuel cell radio to power some tunes? Imagine the scene—you're stuck on a remote island, with your fuel cell experimenter's kit, a solar panel, and an old radio, but your favorite music is played at two in the morning when it's dark and the solar cell won't work! Simple—use the fuel cell as a battery to store energy made during the day for use at night!

You need to find a cheap radio, which takes two or three AA or AAA batteries, i.e., its power supply requirements are between 3 and 4.5 V. Start small, and look for a radio which has minimal power consumption requirements, as this way, your tiny store of hydrogen will last a little longer!

The circuit diagram shown in Figure 14-1 is for our ubiquitous DC-DC converter, which takes our low-voltage power from the fuel cell, and transforms it to a higher voltage output suitable for powering electrical devices—in this case, our radio.

You are going to need to investigate the battery compartment of your radio, and work out where to connect the output from the DC-DC converter. Look at the path that the circuit makes in the battery compartment. It is usual to have a single metal plate covering one battery terminal for where the circuit connects to the battery pack, and a double plate which covers two adjacent battery terminals, where a connection is being made between batteries. The positive terminal will likely be a flat plate and the negative terminal will probably be a spring.

While it may seem like folly to power a radio from an expensive fuel cell, electronics manufacturers are actually looking toward fuel cells to make the next generation of electronics

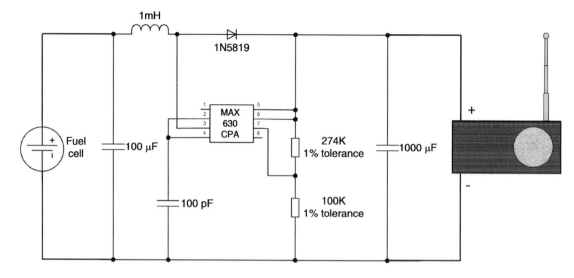

Figure 14-1   *Hydrogen-powered radio schematic.*

Figure 14-2   *Hydrogen-powered radio connected up.*

devices which run for longer. There are limitations with battery technology—batteries are fairly heavy and bulky, and they only last for a certain amount of time. So to get round these limitations, manufacturers are looking at integrating fuel cells into mobile devices, along with replaceable "fuel cartridges" containing hydrogen or methanol acting as a hydrogen carrier.

I've uploaded a short video of this setup to YouTube, which can be accessed at:
www.youtube.com/watch?v=CRhM0ixrmRA

You Will Need

- iPod Shuffle or similar USB MP3 player

- Hydrogen 100 µF capacitor

- 100 pF capacitor

- 1 mH inductor

- 1N5819 diode

- 274 K resistor (1% tolerance)

- 100 K resistor (1% tolerance)

- 100 µF capacitor

- USB socket

- MAX 630 CPA integrated circuit

  or

- DC-DC converter (preassembled) (Fuel Cell Store Part Number: 590709)

- Voller hydrogen generator (Fuel Cell Store Part Number: 72035001)

In this project, we are going to be using our ubiquitous DC-DC converter to power an iPod. Rather than hack the iPod and invalidate the warranty, we are going to get a USB PCB mounted socket, and solder the power outputs from our DC-DC converter. Ensure that you get the wires around the right way as iPods—as with other solid state electrical devices—are polarity sensitive, and

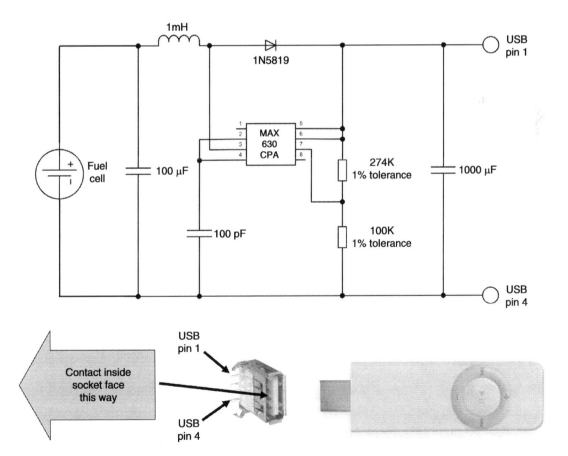

Figure 14-3 *Hydrogen-powered iPod diagram.*

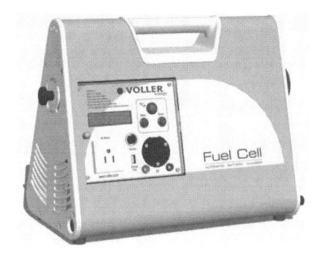

Figure 14-4 *Voller hydrogen generator.*

You will also need to purchase a USB socket, which is used to connect the MP3 player to the DC-DC converter. Again, refer to Figure 14-3, and ensure that you connect the inverter to the socket with the correct polarity, or else you risk damaging your MP3 player.

Another option for powering your portable electronics in remote places is the Voller hydrogen generator (shown in Figure 14-4)—a fuel-cell based generator, which runs on hydrogen, and which provides a USB power output, a 110 V or 230 V (depending on the model) line power output and a 12 V car accessory socket.

The technology is currently out of the reach of the average consumer; however, for certain remote applications the cost is justified, and with advances in fuel cell technology, expect to see the price drop significantly in the next several years.

Many major electronics manufacturers are also investigating the possibility of including fuel cell technology directly as part of their devices. In the next few years, you could be plugging a methanol cartridge into your laptop, portable media player, or mobile phone rather than a battery charger, to provide long-life power for your electronic devices, over and above that possible with present battery technologies.

if you get the wires the wrong way round, you will end up with friedTunes rather than iTunes™.

In this project, we are going to use an iPod shuffle, as they are cheap and have a much lower current draw than the larger hard-disk based iPods. There are a number of other USB MP3 players, and others may work equally well; however, it is impossible to obtain data on the power consumption for all devices.

You will need to assemble the circuit as shown in the diagram, or purchase the ready-assembled printed circuit board from the Fuel Cell Store.

Project 42: Making Hydrogen Bubbles

You Will Need

- Bung
- Thistle tube
- Erlenmeyer flask with side tap
- Swan neck tube
- Acid (e.g., dilute hydrochloric acid)

- Metal (e.g., magnesium ribbon)
- Washing-up liquid, detergent, or soap solution
- Bowl

A fun experiment is to produce some hydrogen—in the example shown in Figure 14-5, we are using the acid-metal method to produce hydrogen relatively quickly. As the hydrogen is produced, and bubbles through the water in the bowl, the hydrogen will form bubbles, which,

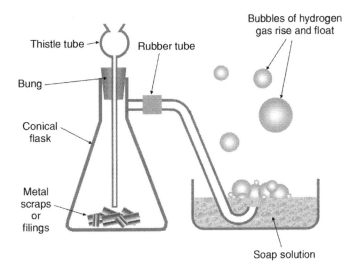

Figure 14-5 *Making hydrogen bubbles.*

being lighter than air, will float. These can then be ignited with a splint. It's another nice, quick and easy, fun experiment, which shows that hydrogen is lighter than air.

Project 43: Exploding Some Hydrogen Balloons

You Will Need

- Latex balloons
- Matches
- Method of generating hydrogen and oxygen (discussed earlier in the book)

Launching exploding hydrogen balloons into the air is a fun way of learning about the science of "stoichiometry." Remember, chemical reactions work best when the reactants are present in the correct quantities. The ideal mixture of hydrogen to oxygen is 2:1—we learn this because the product of the complete combustion of hydrogen is H_2O.

You can use the formula for the volume of a sphere, combined with measurements around the balloon's circumference, to make rough

Figure 14-6 *Exploding hydrogen balloons.*

173

approximations of the volumes of gas you are filling the balloons with.

If we have too much hydrogen present, then when the balloon is ignited, it will have to get oxygen from the surrounding air, providing less of a bang. However, get the mixture of hydrogen and oxygen just right, and a *very* loud bang will ensue.

My former science teacher, Bob Spary, would release balloons filled with the correct mixture—and a long length of burning string as a fuse over our school sports field, while the school soccer team practiced—and the ensuing bang never failed to make them jump out of their skin!

You might like to consider wearing a pair of ear defenders if you have sensitive ears, and, remember,

however you ignite the balloons, make sure you maintain a safe distance!

One possible method of "remote detonation" is to use a simple model rocket igniter taped to the balloon. If you want to read more about model rocket igniter ignition systems, check out my book *50 Model Rocket Projects for the Evil Genius*, McGraw-Hill, 2006.

If you want to watch exploding hydrogen balloons from the comfort of your armchair, take a look at the following cool YouTube videos:

www.youtube.com/watch?v=7UoFNdp0UYg

www.youtube.com/watch?v=MMB2VR0087w

www.youtube.com/watch?v=p3LUHAr12qI

www.youtube.com/watch?v=kknU6cpKWL0

www.youtube.com/watch?v=UUDy9heD6h8

Project 44: Having a Hydrogen Barbecue

You Will Need

- Hydro-Que (Fuel Cell Store P/N: 551005)
- Source of hydrogen

As a conscious environmentalist, no doubt carbon emissions will be your biggest concern when you go out back in the yard to have a grill—not whether the sausages are burnt or not!

Burning gas in the garden is a source of carbon emissions. Patio heaters and gas grills are the enemy of the environmentalist, as they are taking precious fossil fuel resources and squandering them needlessly.

If you are burning wood, worry not if the wood comes from a sustainably managed source, as you

are only putting the carbon dioxide back into the air that the tree consumed when it was growing. However, if you really want to bring your carbon emissions down to zero, consider having a hydrogen-fueled barbecue!

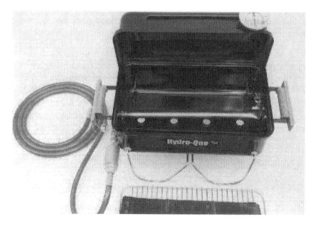

Figure 14-7 *The Hydro-Que hydrogen barbecue. Image courtesy Fuel Cell Store.*

Figure 14-8 *Regulator valve and flash arrestor. Image courtesy Fuel Cell Store.*

The Hydro-Que is a barbecue grill especially modified to accept hydrogen gas as its fuel input.

Remember, at the moment, most hydrogen gas available commercially comes from fossil-fuel resources that are reformed by steam-cracking, so at the moment, the reduced environmental impact of having a hydrogen barbecue is more of a symbolic one than a practical one. However, we can look forward to the day when hydrogen is produced by electrolysis using clean, renewable energy. The Hydro-Que hydrogen barbecue is pictured in Figure 14-7.

Note on the side of the Hydro-Que the gas regulator and flash arrestor. The regulator allows you to vary how much hydrogen is burned, and thus how much heat is produced by the barbecue. The flash arrestor prevents the flame from traveling back down the pipe and igniting the

Figure 14-9 *PURE Energy Centre hydrogen barbecue.*

hydrogen upstream of the barbecue in the tank or hydride.

So now you've eliminated all the carbon emissions from your barbecue, all you have got to worry about is the environmental impact of meat production!

If you're around in the Shetland Islands, check out the PURE Energy Centre, where they have converted a standard barbecue to run on hydrogen with some clever pengineering—check out their fuel cells, and pop by for a sausage!

Chapter 15

Fuel Cell Competitions

If you are interested in taking some of the projects in this book further, there are a number of competitions worldwide open to young people. Competitions are a great way to make use of your skills—hone and refine them, and receive recognition for things that you have learned. Throughout this book, I've flagged up exciting projects which you might like to develop further for science fairs—another great way of showcasing your knowledge and educating your peers about the hydrogen future.

This chapter gives details of all the competitions that are accessible to young people at the time of going to press; however, stay in touch with your National Fuel Cell Association, and keep your eyes peeled for more—as with growing awareness of hydrogen and fuel cell science, there are sure to be new competitions in the future!

High school competitions

International Youth Fuel Cell Competition (IYFCC)

The international youth fuel cell competition is open to young people from around the world. Each country is allowed to select up to three teams to enter the competition—a good starting point would be to contact your National Fuel Cell Association, some addresses of which are listed in the back of this book, to find out about your national entry procedure.

The competition is open to students between the ages of 15 and 18, and an official team consists of two entrants.

Contact details for the IYFCC

Kay Larson: Director

  kay@iyfcc.com

Quinn Larson: Student contact and travel

  quinn@iyfcc.com

Bridget Shannon: Sponsor contact

  bridget@iyfcc.com

Phone: 303-237-3834

Fax: 303-237-7810

Address:

PO Box 4038

Boulder, CO 80306

The great thing about this competition is that it tests your knowledge of fuel cells on so many levels—there is a theoretical element to the competition, in the form of a "quiz bowl," and there are some practical, constructional challenges, where you can apply things that you have learned in this book to build a fuel cell car, and a timepiece powered by fuel cell.

Go to the official IYFCC. website for further information: www.iyfcc.com

Fuel cell car—things to think about

Think about the power-to-weight ratio of your vehicle—the power produced by the supplied fuel cell is fixed—you cannot change that! (But what you can do is ensure that your vehicle is as light as possible!) Also consider the torque and speed produced by your motor—match the power available from your motor as closely as possible with the load, the wheels!

Fuel cell timepiece—things to think about

Think back to the experiments where we produced hydrogen using a PEM fuel cell as an electrolyser, and when we took hydrogen and oxygen and converted them using the fuel cell into electrical power—what did you notice about the reaction when a fixed amount of power was consumed or produced? Was the gas produced at a steady rate or did it vary? Think about the properties of hydrogen—it is lighter than air. Could you trigger a physical action using the lightness of hydrogen gas once a given amount has accumulated? Think about the airships in the chapter on hydrogen transport—think about the formula used to calculate how much hydrogen is required to lift a given weight. Think about electrical power from a fuel cell—could that be used to trigger an action? How could you feed hydrogen to a fuel cell, once a given amount has been produced by an electrolyzer?

Car competition

1. **What components are we required to use on the car?**

   You must use the fuel cell provided at the competition. It may have more power than the one you receive in the practice kit. You must replace the

practice fuel cell with the new fuel cell at the competition. We do not recommend you travel on an airplane with the fuel cell unless you check it in checked baggage. The motor will be the same one provided in the competition kit. You will have access to a new motor at the competition, or you may continue to use the one provided in the competition kit. You may use any other parts included in the practice kit, but you are not required to do so.

2. **Are there any prohibited items?**

   You may not use hydrides or compressed hydrogen storage. You may not use additional motors or fuel cells other than those required. If you are unsure about a component, please contact Kay Larson for approval of questionable components (see the IYFCC Contact Details box).

3. **What are we allowed to bring with us?**

   You may bring a car that is completely built. However, remember that you will need to replace the fuel cell and in doing so, you might need to make some adjustments to your car at the competition.

4. **What will you provide for us to use to make adjustments?**

   There is a picture on the website that includes all of the tools and materials that will be available to you at the competition. Each team will have access to all of the tools and materials. Your car may be completely built, and you might not need to use any of the tools or materials. They are available for you in the event that you need to make major changes, or if you prefer to build your car at the competition using these materials.

5. **How much time will we have to work on our car at the competition?**

   You will have a work day, which is about eight hours. You can divide your time however you choose between working on your car, your timepiece, or studying for the quiz bowl.

Timepiece competition

1. **What components are we required to use on the timepiece?**

   The only component you are required to use is the fuel cell you will receive at the competition.

2. Are there any prohibited items?

You may not use anything that has a timing device in it. The intent of the competition is that the use of hydrogen production and/or storage must be the element that determines the length of time.

3. What do you mean by "something must happen" at exactly two minutes?

At two minutes, there must be some noticeable and predicted event. It could be that something happens at two minutes. It can also be that something stops happening at two minutes.

4. Can we use the fuel cell in electrolyzer mode only if we want to for this competition?

Yes, you can use the fuel cell just as an electrolyzer making hydrogen in this competion.

5. Can we make hydrogen before the two minutes begin?

Yes, you will have 10 minutes to prepare your timepiece before your timing begins. You may use as much of this time as you like to produce hydrogen.

6. How much of the timepiece can we build before we come?

You may bring a completed timepiece with you. Please remember that you will be required to change the fuel cell to a new fuel cell at the competition, and this fuel cell may be more powerful than the one in the practice kit. Be prepared to make adjustments.

7. What will you provide for us to use to make adjustments?

There is a picture on the website that includes all of the tools and materials that will be available to you at the competition. Each team will have access to all of the tools and materials. Your car may be completely built, and you might not need to use any of the tools or materials. They are available for you in the event that you need to make major changes, or if you prefer to build your car at the competition using these materials.

8. How much time will we have to make adjustments at the competition?

You will have a work day, which is about eight hours. You may determine how much of this time

you want to work on your timepiece, your car, or studying for the Quiz Bowl.

Rules and guidelines for the quiz bowl

Purpose

The purpose of the International Youth Fuel Cell Competition Quiz Bowl is to encourage the study of fuel cell technology, and to acknowledge the acceptance of scientific principles as a global educational standard.

Matches

The competition will consist of preliminary rounds and a final round. Teams in the final round will be determined by a double elimination tournament format. Teams will be chosen at random for the preliminary round of questioning, except that teams from the same country will not compete against each other in the preliminary round. After the first round, teams from the same country may compete against one another.

Rules for the quiz bowl

1. Two types of questions will be used: toss-up and bonus questions. A toss-up question may be answered by any member of either team that is playing. The toss-up question must be answered correctly, in order for a team to be offered a bonus question.

2. No team will have more than one opportunity to answer a toss-up question.

3. Questions are either multiple-choice or short answer. The only acceptable answer to a multiple-choice question is the one read by the moderator.

4. Once read in its entirety, a question will not be reread.

5. On toss-up questions, the first player on either team to activate the lock-out buzzer system wins the right

to answer the question. No player may buzz in until *after* the moderator has identified the subject area of the question, e.g., physical science.

6. Before answering the questions, the team member must be verbally recognized by the moderator or scientific judge. (Before the match, this person will be identified.) If not recognized, it is treated as a nonanswer and the moderator will not indicate whether the answer was right or wrong.

7. No consultation is allowed on toss-up questions.

8. Should a player answer a toss-up question before being verbally recognized, or should consultation among any of the team members occur, any answer given does not count (the moderator does not say whether the answer given was correct or incorrect), and the team loses the right to answer the toss-up question. The question is then offered to the opposing team.

9. On a toss-up question, the first answer given is the only one that counts. However, if a student gives both a letter answer and a scientific answer, both parts must be correct.

10. If the answer to a toss-up question is wrong and the question was completely read, the other team is given the opportunity to answer the toss-up question, unless time expires before the second team has buzzed in. The second team is allowed a full 10 seconds to buzz in after the first team has answered incorrectly or has answered without being recognized, unless time expires.

11. The answer to the bonus question must come from the team's captain. Moderators should ignore an answer from anyone but the captain on the bonus question.

Timing rules

12. The match is played until either the time expires or all the toss-up questions have been read. Regional competitions will have two 10-minute halves with a two-minute break. Each half begins with a toss-up question.

13. After reading a toss-up question, the moderator will allow 10 seconds for the two teams to respond before proceeding to the next toss-up question.

Timing begins after the moderator has completed reading the toss-up question.

14. A student who has buzzed in on a toss-up question must answer the question promptly after being verbally recognized by the moderator or scientific judge. After recognizing a student, the moderator will allow for a natural pause (up to three seconds), but if the moderator determines that stalling occurred, the team loses its opportunity to answer the question and it is offered to the opposing team, if eligible.

15. After a team member has answered a toss-up question correctly, the team is given the opportunity to answer a bonus question. The team will have 30 seconds to begin to give its answer to the bonus question. Consultation among team members is allowed on bonus questions.

16. On a bonus question, the signal "5 SECONDS" will be given by the timekeeper after 25 seconds of the allowed 30 seconds have expired. Additionally, the timekeeper will indicate the end of the 30-second bonus period by saying "TIME." If the team captain has not begun the response before the timekeeper calls "TIME," the answer does not count. If the team captain has begun the response, he or she may complete the answer.

Scoring

17. Toss-up questions are worth four points, and bonus questions are worth 10 points.

18. If a toss-up question is interrupted, the student recognized, and the answer correct, the team will receive four points. If the answer is incorrect, four points are added to the opposing team's score, the question is reread in its entirety, and the opposing team has an opportunity to answer the toss-up question with the chance to answer the bonus question if correct.

19. The double interrupt—if a toss-up question is interrupted, the student has been verbally recognized and the answer is incorrect, four points are added to the opposing team's score. The question is then reread in its entirety. However, if a student on the opposing team interrupts the rereading of the question, the player is verbally

recognized and gives an incorrect answer. Four points are added to the other team's score. The moderator will give the correct answer and move on to the next toss-up question.

20. If the moderator inadvertently gives an answer to a toss-up question without giving either team a chance to respond, no points are awarded and the moderator goes on to the next toss-up question.

21. If a toss-up question is interrupted, the student is *not* recognized and blurts out an answer, the result is a nonanswer. No penalty points are awarded to the opposing team. The moderator will not indicate whether the answer was right or wrong and the question is reread in its entirety to the opposing team.

22. If the moderator inadvertently gives the answer to a toss-up question before allowing the second team to respond (after an incorrect answer, or an answer given without the team member having been recognized), the next toss-up question will be read to the second team in place of the inadvertently answered question.

Summary of Scoring:

Type of Question	Points Awarded
Toss-up	+4 points, and eligible for bonus
Bonus	+10 points
Incorrectly answered	+4 points to opposing team
Interrupted toss-up	+4 points to opposing team
Unrecognized and interrupted toss-up	+0 points
Unrecognized toss-up	+0 points

Challenges

23. Challenges to questions and responses will be permitted. Team members may also challenge a ruling, the score, or a protocol issue. A challenge can only be made by a team member who is actively competing. A challenge may not be made by the coach, alternate, or by anyone else in the audience.

All challenges must be made before the next question is begun. The scientific judge and the moderator may consult during the match regarding responses. All decisions made by the judges are final.

24. Should a question arise during a competition, the competition and the clock will be stopped until the question is resolved. Once the question has been resolved, the match will continue from that point. Should the moderator decide that some time was lost due to the interruption, the moderator has the right to put the appropriate amount of time back on the clock.

When time runs out

25. If the question has been completely read, but neither team has buzzed in, the game or half is over.

26. If the question has been completely read, a student has buzzed in and is recognized before answering, and gives a correct answer, the team gets to answer the bonus question. The half or game is then over.

27. If the question has been completely read, a student has buzzed in and is recognized before answering, but gives the wrong answer or answers before being verbally recognized, the game or half is over.

28. If the question has been completely read, a student has buzzed in and time is called before the student has been recognized, the moderator or scientific judge will verbally recognize the student. If the student gives a correct answer, the team gets to answer the bonus question. If an incorrect answer is given or the student answers before being verbally recognized, the game or half is over.

If the question has NOT been read completely before time runs out:

29. If the question has not been completely read by the moderator and neither team has buzzed in (interrupted), the game or half is over.

30. If a team member buzzes in before time is called, interrupting the reading of the question, is verbally recognized, and answers the question correctly, the team gets to answer the bonus question. The half or game is then over.

31. If a team member buzzes in before time is called, interrupting the reading of the question, and is

verbally recognized, but answers the question incorrectly, penalty points are awarded, the question is reread for the other team, which is then given the chance to answer both the toss-up question and, if correct, the bonus question before the contest or half is over.

32. If a team member buzzes in before time is called, is not verbally recognized, and blurts out the answer, the answer is not accepted but no penalty points are awarded. The question is read in its entirety for the other team which, if it answers correctly, is also given a chance to answer the bonus question before the contest or half ends.

Rules for entering the final round

33. In the event that the required number of teams from each division is not clearly identifiable [resolution is necessary only between teams tied for last position(s) to advance to the single or double elimination], a tie-break procedure in the following order will be used:

    a) Head to head won/loss record

    b) Fewest losses

    c) If two teams are still tied, there will be a five toss-up question run-off (interrupt penalty in effect). No bonus questions will be used during this segment of the competition. If still tied, another five toss-up question run-off will be used, etc. until the advancing team is determined.

    d) If more than two teams are tied, each team, in separate rooms, will be given a series of 10 toss-up questions (no bonus questions will be used during this segment of the competition). The usual 10 seconds will be allowed for a competitor to buzz in after the question is completely read. There are no interrupt penalties but also no reason to interrupt since all 10 questions will be read. Scoring will be based on the number of questions right minus the number wrong. If two or more teams are still tied, procedure (iii) or (iv), as appropriate, will be used until the advancing teams are determined.
    [. . .]

Rules for the end of a single or double elimination match

51. If the score is tied in a single or double elimination match at the end of the regulation time period, a series of five toss-up questions will be used to break the tie. Interrupt penalties are in effect. Round robin matches may end in a tie, as explained in rule 15 above.

Miscellaneous science bowl rules

52. No one in the audience may communicate with participants during the match; communication will result in a warning, and if the problem persists could result in an ejection from the competition room.

53. If someone in the audience shouts out an answer, the question will be thrown out (as will the person), and the moderator will proceed to the next question.

54. Prior to each match, the two team coaches will introduce themselves to each other and will sit together in the back row of the competition room.

55. No notes may be brought to the competition. Nothing may be written before the clock starts. Scratch paper will be provided at the beginning of each match, and collected at half-time and at the conclusion of the match.

56. Calculators are not permitted.

57. Members of the audience, including the coaches, will not write down the questions or answers the moderator reads, or use any electronic recording or transmitting device, including digital cameras during the match. At the nationals, coaches will be provided with a team score sheet to track the number of questions answered by each individual student on their team. No one else in the competition room is permitted to make notes of any kind during the active competition. If this occurs, the individual(s) will be asked to leave the competition room.

Hydrogen student design contest

The hydrogen student design contest is an annual competition that takes students with a knowledge of fuel cell technology, and gets them to apply this knowledge to a design problem. The contest focuses on practical applications utilizing hydrogen and fuel cells.

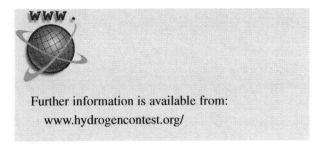

Further information is available from:
www.hydrogencontest.org/

The rules vary every year as different challenges are set annually.

Department of Energy National Science Bowl (United States)

The DOE National Science Bowl includes an event in the finals—the "Hydrogen Fuel Cell Model Car Challenge." For this, apply principles that you have learned from the chapter on hydrogen transport (Chapter 13).

More info about the DOE Science Bowl can be found at:

www.scied.science.doe.gov/nsb/index.html

Fuel Cell Box (Germany)

A national competition to raise awareness about fuel cell technology among schoolchildren in Germany called "Fuel Cell Box" may be of interest to you. Further information, in German, is available from www.fuelcellbox.de.

Appendix A

Everything You Wanted to Know About Hydrogen ... and Then Some

Chemical series	Nonmetals
Group	Period, Block 1, 1, s
Electron configuration	1s1
Electrons per shell	1
Appearance	Colorless

Gas* (m³)	Liquid (liter)	Weight (kg)
1	1.163	0.0898
0.856	1	0.0709
12.126	14.104	1

*m³ at 981 mbar and 15° C

Data	Value
Density	0.08988 kg/nm^3
Upper heating value	12.745 MJ/nm^3
Lower heating value	10.783 MJ/nm^3
Ignition energy	0.02 MJ
Ignition temperature	520° C
Lower ignition level	(gas concentration in air) 4.1 Vol%
Upper ignition level	(gas concentration in air) 72.5 Vol%
Flame rate	2.7 m/s
Melting point	14.01 K (−259.14° C, −434.45° F)

Data	Value
Boiling point	20.28 K (−252.87° C, −423.17° F)
Triple point	13.8033 K, 7.042 kPa
Critical point	32.97 K, 1.293 MPa
Heat of fusion (H_2)	$0.117 \text{ kJ·mol}^{-1}$
Heat of vaporization (H_2)	$0.904 \text{ kJ·mol}^{-1}$
Heat capacity (25° C) (H_2)	$28.836 \text{ J·mol}^{-1}\text{·K}^{-1}$
Thermal conductivity (300 K)	$180.5 \text{ mW·m}^{-1}\text{·K}^{-1}$
Speed of sound	(gas, 27° C) 1310 m/s
CAS registry number	1333-74-0 (H_2)
Crystal structure	hexagonal
Oxidation states	1, −1
Electronegativity	2.20 (Pauling scale)
Ionization energies	1st: 1312.0 kJ/mol
Atomic radius	25 pm
Atomic radius (calc.)	53 pm (Bohr radius)
Covalent radius	37 pm
Van der Waals radius	120 pm

Hydrogen density compared to other fuels:

	Hydrogen	Gasoline	Diesel	Natural gas	Methanol
Density (kg/l)	0.0000898	0.702	0.855	0.00071	0.799
Density (kg/m^3)	0.0898	702	855	0.71	799
Energy density (MJ/kg)	120	42.7	41.9	50.4	19.9
Energy density (MJ/l)	0.01006	31.2	36.5	0.00361	15.9
Energy density (MJ/m^3)	10783	31200	36500	36.1	18000
Energy density (kWh/kg)	33.3	11.86	11.64	14	5.53
Energy density (kWh/m^3)	2.79	8666.67	10138.88	10.02	4420

All numbers are at lower heating value and at atmospheric pressure and normal temperature.

Fuel Cell Acronym Buster

In the world of fuel cells, they love to abbreviate things! You will come across some of these acronyms in this book, while others you will find are used in conversation and other books when looking at fuel cells and their applications. This handy little guide should help to debunk the TLAs (Three Letter Acronyms) into something a little more intelligible.

AFC alkaline fuel cell

AFV alternative fuel vehicle

BEV battery electric vehicle

CH₃OH methanol

CH₃CH₂OH ethanol

CHP combined heat and power

CNG compressed natural gas

CO carbon monoxide

CO₂ carbon dioxide

DMFC direct methanol fuel cell

FC fuel cell

FCEV fuel cell electric vehicle

GDE gas diffusion electrode (See GDM)

GDM gas diffusion media/membrane (See GDE)

FC fuel cell

FCEV fuel cell electric vehicle

GHG greenhouse gas

H₂ hydrogen

HC hydrocarbons

ICE internal combustion engine

LPG liquefied petroleum gas

MCFC molten carbonate fuel cell

MEA membrane electrode assembly

MeOH methanol

NABH₄ sodium borohydride or sodium tetrahydroborate

NOₓ oxides of nitrogen

O₂ oxygen

PAFC phosphoric acid fuel cell

PEFM polymer electrolyte fuel cell

PEM proton exchange membrane or polymer electrolyte membrane

PEMFC PEM (see above) fuel cell

Pd palladium

Pt platinum

SOFC solid oxide fuel cell

URFC unitized regenerative fuel cell

Fuel Cell Associations

If you are interested in finding out what schools, colleges, and universities run courses in your area, or if you are interested in work experience and want to find a suitable company, or if you are simply interested in what's happening with hydrogen in your area, contact your local hydrogen association, where they will be more than happy to give you information to further your interest in hydrogen and fuel cells.

Europe

European Hydrogen Association

Gulledelle 98

1200 Bruxelles

Belgium

Tel: 32 2 763 25 61

Fax: 32 2 772 50 44

E-mail: info@h2euro.org

www.h2euro.org

French Hydrogen Association

Association Française de l'Hydrogène

28, rue Saint-Dominique

75007 Paris

Tel: 33 (0)1 53 59 02 11

Fax: 33 (0)1 45 55 40 33

E-mail: info@afh2.org

www.afh2.org/

Fuel Cell Europe

Fuel Cell Europe

World Fuel Cell Council e.V.

Frankfurter Strasse 10-14,

D-65760 Eschborn

Germany

www.fuelcelleurope.org

Fuel Cells UK

Synnogy Ltd

1 Aldwincle Road

Thorpe Waterville

Northants NN14 3ED

Tel: 44 (0)1832 720007

E-mail: info@fuelcellsuk.org

www.fuelcellsuk.org

German Fuel Cell Association

Unter den Eichen 87

12205 Berlin

Germany

Tel: (49-700) 49376 835 (hydrotel)

Fax: (49-700) 49376 329 (hydrofax)

E-mail: h2@dwv-info.de

Norwegian Hydrogen Forum

radix.ife.no/nhf/

Polish Hydrogen and Fuel Cell Association

Polskie Stowarzyszenie Wodoru i Ogniw
Paliwowych

molenda@uci.agh.edu.pl or

hydrogen@agh.edu.pl

E-mail: www.hydrogen.edu.pl

Scottish Hydrogen and Fuel Cell Association

Brunel Building

James Watt Avenue

Scottish Enterprise Technology Park

East Kilbride

G75 0QD

Scotland

Tel: 44 7949 965 908

www.shfca.org.uk

Spanish Hydrogen Association

Asociación Española del Hidrógeno

Head Office:

c/o Isaac Newton, 1

PTM. Tres Cantos

28760 Madrid

Spain

CIF: G-83319731

Tel: 34.654.80.20.50

Fax 34.91.771.0854

E-mail: info@aeh2.org (about AeH2)

E-mail: aeh.web@ariema.com (about this website)

aeh2.org/en/index.htm

America

American Hydrogen Association

2350 W. Shangri La

Phoenix, AZ 85028

Tel: (602) 328-4238

Fuel Cells Texas

c/o Good Company Associates Inc.

816 Congress Avenue, Ste. 1400

Austin, TX 78701

Tel: (512) 279-0750

Fax: (512) 279-0760

www.fuelcellstexas.org/

Hydrogen and Fuel Cells Canada

4250 Wesbrook Mall

Vancouver, B.C. V6T 1W5

Tel: 604-822-9178

Fax: 604-822-8106

E-mail: admin@h2fcc.ca

National Hydrogen Association

Main Office:

1800 M St NW, Suite 300 North

Washington, DC 20036-5802

U.S.A.

Tel: 202-223-5547

Fax: 202-223-5537

Western U.S. Office

35 Seacape Drive

Sausalito, CA 94965

Tel: 415-381-7225

Fax: 415-381-7234

E-mail:infoWestUS@ttcorp.com

www.hydrogenassociation.org/

North Carolina Sustainable Energy Association

NCSEA

PO Box 6465

Raleigh, NC 27628-6465

Tel: 919-832-7601

E-mail: ncsea@mindspring.com

www.ncsustainableenergy.org/

Asia

**China Association for
Hydrogen Energy (CAHE)**

Room 710

No. 86 Xueyuan Nanlu

Beijing

100081

China

Tel: (86-10) 62180145

Fax: (86-10) 62180142

Oceania

National Hydrogen Association of Australia

13 Mount Huon CCT

Glen Alpine

Campbelltown

New South Wales 2560

Australia

Tel: 61 2 9603 7967

Fax: 61 2 9603 8861

www.hydrogen.org.au/nhaa/

Epilogue

What does the future hold for fuel cells? Why don't I ask you the same question? You're the one with this book in your hand, a head full of great ideas, and a desk full of fuel cell equipment—go and create the next generation of sustainably powered inventions, gadgets, and gizmos.

Fuel cells have already undeniably helped to shape the world—they've helped us in our quest to conquer the mysteries of space, by producing electricity to power space capsules, while providing water for astronauts to drink. They've also helped us subjugate the oceans, allowing submarines to stay underwater for weeks on end.

The challenge now, as the technology finds more applications, is to vanquish our addiction to carbon-based fuels. It's a bad habit which we've acquired over the past couple of centuries, but one that we need to kick quickly, as we have limited resources, and are faced by the problems of environmental degradation, photochemical smog, and the threat of climate change—all nasty side-effects of an infatuation with coal and oil.

The fuel cell will enable a cleaner economy—based on hydrogen, not carbon—facilitating a transition to greener technologies and new models for providing power and energy services. Fuel cell technology is going to release consumers of power from the shackles, and inefficiency of centralized thermal generation (where the bulk of the input energy is wasted as heat) to new modes of decentralized generation, where the heat produced from an energy conversion process is used to good effect—to heat your home or to warm your community's swimming pool, rather than being wasted into the atmosphere in plumes of smoke.

It's also going to enable us to use increased amounts of renewable energy as part of our energy generation portfolio—we can only produce solar power when the sun shines, and wind power when the wind blows, but we need mechanisms to store it when it's being produced but not used. Again, fuel cells are very helpful here as they enable us to "mop up" spare energy when a surplus is being produced, and store it as hydrogen later to be used as a transport fuel, or converted into electricity.

At the PURE Energy Centre, Shetland, UK and the HARI project, West Beacon Farm, UK, they are doing just that—learning by experiment, just as you're doing with this book, how to use hydrogen produced from renewables to meet their energy needs.

The technology is developing every day, and every person that joins the army of scientists, engineers, technologists, and product designers working to bring fuel cells onto the market and into the mainstream, helps the movement gain a little bit of momentum.

The future is bright for hydrogen—Hjaltland Housing Association, along with the PURE Energy Centre and Fuel Cells Scotland, have released plans for the world's first off-grid fuel cell houses. The houses will have wind turbines in their yard to produce power, and will turn this power into hydrogen which can be used to meet the homes' energy needs.

This sort of development provides a blueprint for clean homes around the world, and could be coming to a block near you.

*Architects drawing for Hjaltland Housing Association
fuel cell homes. Image courtesy PURE Energy Centre*

Don't let people put you off fuel cells by saying
that they are expensive and will never be affordable—
remember that big TV your Pa wanted a few years
ago, but was a touch on the pricey side . . . that he
bought last week because it was a third of the
price, or that games console that a couple of
Christmases ago was at a premium price, but now
has dropped in price and is a bit more affordable?
It's all proof that there is a relentless drive to
engineer the cost out of technology to make it
accessible to a wider marketplace. Fifty years ago,
it would have been inconceivable for every one of
the world's large corporations to have a computer.
Today, they are on the desk of every employee,
and furthermore—with initiatives like "One Laptop
Per Child"—the technology is even accessible to
the developing world. The same will happen with
fuel cells. Problems are surmounted, cheaper ways
to make better products are found, and the
products become ubiquitous.

So watch out . . . fuel cells may be coming to a
home near you. In fact . . . they already have!

Index

Page numbers for figures and tables are given in **bold**

Made in the USA
Lexington, KY
02 June 2016